Marshall Monroe Kirkman

**Railway Service**

Trains And Stations

Marshall Monroe Kirkman

**Railway Service**
*Trains And Stations*

ISBN/EAN: 9783744724159

Printed in Europe, USA, Canada, Australia, Japan

Cover: Foto ©berggeist007 / pixelio.de

More available books at **www.hansebooks.com**

# Service

## Trains and Stations.

DESCRIBING THE MANNER OF OPERATING TRAINS, AND THE DUTIES
OF TRAIN AND STATION OFFICIALS.

BY

MARSHALL M. KIRKMAN.

PUBLISHED BY
THE RAILROAD GAZETTE, No. 73 BROADWAY, NEW YORK.
1878.

# PREFACE.

The physical life of a railroad, is known in its entirety by comparatively few, but it is nevertheless true that a comprehensive knowledge of the laws that regulate and direct that life is essential to every railway man of any prominence, or that hopes one day to become prominent. Many men connected with our railroads do not, however, possess the facilities necessary for acquiring this knowledge. This book, it is hoped, will be of assistance to all such. It treats of the composition and movement of Railway Trains and the Laws governing the same, including an exposition of the duties of Train and Stationmen.

In pursuing my inquiries in reference to the subject, I have had occasion to examine into the regulations of all the great companies of the United States and England, so far as they affect train and station service. The results of these investigations are embodied herein.

Chicago, *May*, 1878.

# TABLE OF CONTENTS.

                                          PAGE.

Preface . . . . . . . . . . . iii

## CHAPTER I.

The mysteries that underlie the organization and movement of trains — The block system — Manipulation of trains upon English roads — The force employed — The collection of fares — The schedule by which regular trains are moved — Movement of trains by telegraphic orders — The protection of trains. . . . . 1

## CHAPTER II.

Individuality of railroad companies — Dissimilarity of the signals in use upon different roads— The danger such dissimilarity renders possible— Want of uniformity in the rules and regulations governing different roads — Intelligent discrimination exercised by trainmen — The conservatism of trainmen — The regulations partake of the character of the men introducing them — Some of the differences observable in the rules and regulations of different railroads — No uniformity in the telegraph department— Lack of completeness and thoroughness in framing the rules and regulations— The wonderful phraseology of trainmen — Phraseology peculiar to English roads. . . . . . . 27

## CHAPTER III.

Explanation of some of the technical terms in use in connection with the train and station service of a railway company. . . . . . . . . . . 51

## CHAPTER IV.

Plan pursued in arranging and compiling the rules and regulations. . . . . . . . . . 65

## CHAPTER V.

Signals required by railway companies — Train signals — Enginemen's signals — Conductors' signals by bell cord — Signals by hand — Regulations governing the use of signals. . . . . . . . 69

## CHAPTER VI.

Classes and grades of trains — Rights of trains — How to protect trains when standing upon the main track, or when the track is obstructed — When trains break in two — Trains running with care — Trains must stop — Trains meeting or passing each other — Trains approaching stations — Trains following other trains — Keeping off the time of other trains — Delayed trains — Extra trains — Construction and wood trains — Wild trains — The speed of trains — Directions applicable only to double track lines — Third track or middle sidings — Coupling cars — Miscellaneous orders relative to trains — The track — Movement of trains by telegraph. . . . . . . . . . 81

## CHAPTER VII.

General instructions to conductors — Passenger conductors — Freight conductors — General instructions to brakemen — Passenger brakemen — Freight brakemen — Train and station baggagemen — Enginemen — Firemen — Inspectors of engines — Yard masters. . 147

## CHAPTER VIII.

Telegraph Operators — Telegraph Repairers. . . 188

## CHAPTER IX.

Agents — Rules referring to the passenger traffic — Freight regulations — Directions to agents — Receiving freight for shipment — Receipting for freight — Releases — Loading and unloading freight — Care must be exercised in loading freight — Delivery of freight — Freight from and to stations at which there are no agents — Waybilling freight — Directions to agents in reference to sealing cars containing freight — Miscellaneous rules for freight agents — Directions to agents in reference to fuel — To switches — To trains and cars — General directions to agents. . . . . . . . 195

## CHAPTER X.

General Instructions. . . . . . . 221

## CHAPTER XI.

Regulations of the Austrian railways governing the passenger service . , . . . . . 228

## CHAPTER XII.

A chapter devoted to the rules and regulations of the great English roads — General Regulations — Conditions under which persons are admitted to the service — Security — Privileges — Compensation, etc. — The Uniforms required and the regulations incident thereto — General regulations for working the absolute block system on a double track road. . . . . 238

# RAILWAY SERVICE:

## TRAINS AND STATIONS.

---

### CHAPTER I.

#### THE MYSTERIES THAT UNDERLIE THE ORGANIZATION AND MOVEMENT OF TRAINS.

The manipulation of trains never ceases to be a subject of wonder and speculation to railway men.

To the great bulk of them the secrets that envelope the construction of the schedule by which trains are moved are profound and impenetrable.

How the officials are able to control the labyrinth of moving trains, how watch them as they wind in and out like the figures upon a chessboard, how adjust so nicely the time of their arrival at meeting and passing points, how keep them all in motion, regulate their speed and give to each the exact consideration its importance merits, are questions that but few railway men can understand. They know that there is hidden away somewhere in the dark unoccupied recesses of the Superintendent's apartments a mysterious chart, whereon at intervals he works. It is upon this that he fixes the character, speed, and stopping places of trains, here he notes where they shall meet or pass each other, not forgetting the time they shall start, nor the hour they shall reach their destination. They have had surreptitious glimpses of this wonderful chart through partly closed doors, but their view has been obstructed and their mental processes deadened by the form and austere presence of the Superintendent as he paced the room with measured stride, or bent over his work, pencil in hand, with absent air and corrugated brow, like one who sought in vain the solution of some difficult problem. They have noted with awe the hieroglyphics pregnant with meaning that cover the broad white surface of the mysterious chart, the stations printed in big black letters of varying size and type, and seeming to derive a fictitious importance from that fact; the broad lines of different color that traverse its face laterally and at right angles. Nor have they failed to note and comment upon the faint irregular lines drawn with tremulous hand, here and there, without method or object, apparently, lines seemingly taking their rise in space and ending in space, feeble, inconsequential, indefinite, like disconnected dreams or half completed thoughts.

But while they know or surmise that these faint, irregular, half-obliterated lines forecast moving trains, that they represent organized harmonious action, that each line is a fully developed, completed idea, they do not know how these ideas, clothed in the symbolical language in which they see them spread upon the chart, are to be subsequently arranged and grouped, how condensed into the simple form they present in the printed time table or schedule, which they have carried in their pockets for years.

While any of us may without much labor become acquainted with the charts that the Superintendent uses in constructing his table of trains, still we can not without study, and long association with his duties and responsibilities, understand all the nice distinctions that govern him in his work. Nevertheless, each schedule presents many features that seldom, if ever, change; certain trains become in time like the staple articles that a grocer is compelled to keep, whether he derives profit therefrom or not; their abandonment can not be contemplated, and the most trivial changes in their organization or time may precipitate upon the hapless Superintendent the indignation of an outraged community; this indignation at once finds utterance and relief in long petitions, sarcastic newspaper articles, crowded mass meetings, and waiting committees.

Aside from the staple features noticeable in the list of trains, the probable amount of business that will offer, its source, and the direction it will take, have to be carefully considered in constructing the schedule. But these calculations, made from time to time as new schedules are constructed, may be said to have reference only to the freight traffic, and the number of trains required to do the business with reasonable expedition and economy.

The number of passenger trains employed upon our roads are seldom, if ever, reduced. On the contrary, new trains are added at long intervals as the country developes and the business of the line increases. The various passenger trains move back and forth on a fixed course, year after year, with the dull monotony of an ever swinging pendulum; each train has a name and character along the route it follows, and people speak of it as they do of their recurring crops.

More or less of the freight trains that are operated may also be classed with the staple articles; a certain number, varying with the size and character of the road, are necessary to do its business; like the passenger trains they will present at certain seasons of the year an exceedingly meagre, if not beggarly, appearance, but they are necessary to the convenience of the community and an expeditious conduct of the varying business that is offering, and so they escape the inevitable reduction that overtakes unproductiveness or extravagance in other branches of the business.

Many other things have to be considered and provided for in arranging the schedule. It is desirable there should be close connections at various junctions with other roads. It is this phase of the subject that tries the patience and ingenuity of the official. While no one of us, perhaps, but has felt gratified at being able to make easy and swift connection at some junction on our route, not all of us have stopped to realize that the (to us) propitious conjunction of circumstances was not the result of chance, but of much contention, of many long and angry communications, much bitterness of feeling, succeeded by many agreements and counter agreements, these in turn finding, eventually, definite and final solution in some happily devised compromise that represented approximately the rights of each company interested. It is sometimes the case that connection with other lines has only to be made at one end of the road. It is thus of the greatest consequence to the management that trains going in a certain direction should reach their destination at a particular hour, but when these trains shall start upon their return journey is purely a local question, to be considered only in the relation it bears to the other local questions of the company.

The experience and skill required to move trains with economy and safety upon a single track is infinitely greater than that required where two or more tracks are available. Indeed, the ability required to manipulate trains successfully may be said to be in the inverse ratio to the number of tracks upon which they are moved. Upon a double track road there is no necessity of providing meeting places for trains. Where there is but one track, this is of the greatest consequence, as trains can only pass each other at those points where adequate sidings have been provided. The sidings at a particular station may be of sufficient length to enable passenger trains to meet and pass, but not adequate to the passage of freight trains. The nicest calculations have, therefore, to be made to so arrange the movement of trains that the meeting or passing places may occur at points where the accommodations are adequate.

When three tracks are available for the movement of trains, the special provision required upon a double track line to enable trains moving in the same direction to pass each other without delay or inconvenience, is greatly lessened, if not entirely obviated. When it is necessary for a train to pass another, where three tracks are employed, the forward train pursues its way at reduced speed, upon the third or intermediate track, while waiting for the fast train in its rear to overtake and pass it, before it can re-occupy the main track. It will of course sometimes happen that a particular section of the third track will be required for use simultaneously by slow trains moving in opposite directions. When this is the case, the opposing trains will be compelled to wait until one of them can with safety re-occupy the main track.

When four tracks are employed, the manipulation of trains becomes still more simple. It is no longer intricate or elaborate. It is simply a matter of calculation, affording abundant scope, doubtless, for the exercise of good judgment and tact, but not requiring the elaborate experience and skill necessary where the facilities are more restricted. Trains of the same class, of equal or approximate grade, follow each other in endless succession, and only the local or accommodation trains are required, at long and comparatively infrequent intervals, to give way to faster trains of a higher order, but of the same grade.

Upon a four track railway the danger to life and property may be said to have reached the minimum, while the facility of business and the economy of operation have reached the maximum.

When separate tracks have been provided for moving trains in opposite directions, it would seem as if life and property were surrounded with every reasonable safeguard against the danger to be apprehended from colliding trains, but it is undoubtedly true that disaster perpetually menaces trains following each other in quick succession at a high rate of speed.

While the results to be apprehended from a train being run into from the rear do not at first sight seem likely to be as disastrous as they would be from trains colliding while moving in opposite directions, yet a moment's reflection makes it apparent that the danger to life is, under certain circumstances, really much greater in the former, than in the latter case.[1]

### THE BLOCK SYSTEM.

This may be described as a system devised to secure the expeditious movement of trains upon a road possessing two or more tracks, without jeopardizing life or property.

---

1. "Any one who has examined our reports of train accidents, will have observed that about one-fifth of all those reported are rear collisions which would be impossible when working with the block system."—*Railroad Gazette.*

Under its workings the track of a road is cut up into short sections of a few miles in length, called blocks.

Under what is termed the "absolute" system, not more than one train is allowed upon a block at the same time, consequently a collision is impossible so long as trains remain upon the track.

The "permissive" system allows more trains than one to move upon a block at the same time, under certain circumstances, but it provides specifically for notifying each train that enters a block, whether such block is unoccupied or not.

When a train passes off from a block, it is noted by the operator, and the fact instantly telegraphed to the signalman at the opposite end of the section that is vacated; the track thus becomes free for the use of any following train.

Until the receipt of this notice no train is permitted to enter the block, under the "absolute" system.

Under the "permissive" system, certain trains would be allowed to enter after having been notified that the block was already occupied.

The block system makes provision for keeping the officials of a train advised when the track is obstructed by preceding trains; the danger of trains being run down is thus rendered practically impossible.

The system is highly esteemed abroad, and is in limited use in this country.

The enormous cost of the appliances necessary to the operation of the block system, and the great expense attendant upon its workings, may be said to practically prohibit its general introduction in the United States for the present. The wealthier companies will in time adopt it, and it will be introduced upon isolated sections of road where the business is so great as to endanger the safety of trains operated under the ordinary rules.

The system may be said to be indispensable where the business of a company is such as to require that trains should succeed each other at intervals of only a few minutes.

It is relatively of much greater importance to a company with two tracks than one with double that number.

The danger of trains running into each other can not be so great with four tracks as with two, for the reason that while the number moving in the same direction upon any one track may be as great upon one road as the other, still the trains that succeed each other upon one line will all be moving at a comparatively uniform rate of speed, while upon the other they will vary from fifteen to sixty miles per hour.

Besides this, while it is not improbable that

the freight traffic of a road may increase proportionately with the number of its tracks, still the number of passenger trains required is not likely to be similarly affected and thus the tracks allotted to such trains are comparatively idle.

There is no doubt but what the uniform rate of speed pursued by trains following each other upon a four track road affords a protection impossible upon roads where a less number are available, but in the event a train is delayed or one or more of its cars become detached, the danger is just as great upon a four track road as upon one in which only two are employed, supposing the business to be proportionately the same. It is absolutely essential under such circumstances that a following train should be warned that a train or portion of a train is in its immediate front.

The block system takes cognizance of every attending circumstance, and if, under its workings, a train were to break in two, and the forward part continue on its course ignorant of the fact, the loss would instantly be observed by the operator and the block would not be freely opened to succeeding trains until the facts were fully investigated.

## MANIPULATION OF TRAINS UPON ENGLISH ROADS — THE FORCE EMPLOYED — THE COLLECTION OF FARES.

The great English roads are all operated under the block system, or what may be termed a modification of such system. Each line is thickly dotted with signal houses and their attendant appliances. The great bulk of the rules and regulations under which our trains are operated have, therefore, no relevance with them.

While they provide schedules as we do, yet the trains are constantly guarded and protected by the multitude of signalmen scattered along the line.

These men are ubiquitous; trains move or remain stationary as they direct; they approach or remain away from stations at their beck or nod, and when a train has reached a station it departs or not as the signals indicate. So that while trains may be behind time, or may not be recognized by the schedule, they still pursue their way with undiminished speed so long as the signals in their front indicate the track to be clear.[1]

---

[1]. A very full description of the workings of the block system is embraced in a succeeding chapter. This description is taken from the rules and regulations of the English roads operated in accordance with the Clearing House Standard.

The trains manipulated under the eye of the signalmen, of course require double tracks upon which to move.

Upon single track roads in Great Britain the great utility of the telegraph in connection with the movement of trains is practically unknown, and in that respect our system of management is immeasurably superior to theirs.

The duties of the conductor abroad are exceedingly diverse. He may be said to be the creature of innumerable circumstances. Frequently without an assistant on board the train, he is expected to assist in its protection; perform the duties of a brakeman; act as express-messenger, baggage-master and attendant. Nominally in control of the train when upon the line, his authority vanishes upon its arrival at a station. He assists passengers in entering and leaving the cars, but their fares are collected by another.[1]

The elaborate force which mans our passenger trains is unknown in England. There the force consists of a guard (conductor), as intimated above.

He does not always have an assistant.

---

1. "Should a guard have reason to suppose any person is without a ticket, or not in the right carriage, he is to request the party to show him his ticket, not with a view to receive it from him, but to satisfy himself that every passenger has a proper one. He is under no circumstances to receive money on account of the company."—*Regulations English Roads.*

The head guard has charge of the train, and its passengers, baggage and express matter.

The assistant guard has a box in one of the cars or vans; he signals the train in case of danger, attends to the brake, and performs such other duties as he may be able.

In lieu of these men we usually have a conductor, express-messenger, baggage-man and two brakemen. Our station service is, however, conducted with a much less force than theirs.

Their apparent extravagance in this respect is explained in part by the fact that the rules requiring passengers to purchase tickets before entering the cars are rigidly enforced by them. The outlay is, therefore, not an extravagance.

In connection with this subject of passenger fares and their payment, the regulation of the Austrian roads, contained elsewhere herein, that permits and directs the officials of a company to impose a fine upon passengers who neglect to purchase tickets, or claim that they did not have time to purchase them, is interesting and instructive. The laws of England governing the time and manner of paying passenger fares are also exceedingly strict.[1]

---

[1]. "Under the provisions of the acts relating to this railway, any person who shall travel or attempt to travel in any carriage used on the railway, without having previously paid his fare, and with intent to avoid payment thereof, or who, having paid his fare for a certain distance, shall knowingly and wil-

## THE SCHEDULE BY WHICH REGULAR TRAINS ARE OPERATED.

An economical management of railway property requires that the printed schedule, in accordance with which trains are operated, should provide only for the minimum number required to do the business of the road. The schedule specifies the precise minute each train shall start upon its journey, the time of its arrival at the various stations and sidings, and, finally, the hour it shall reach its destination.

A glance at the table tells us where trains meet or pass each other, such places being indicated with startling distinctness by great fat dropsical looking figures that instantly engage the eye, and arrest the attention of the most superficial observer; this is doubtless why they are used, and it is very likely for the same

fully proceed in any such carriage beyond such distance without previously paying the additional fare for the additional distance, and with intent to avoid payment thereof, or who shall knowingly and wilfully refuse or neglect, on arriving at the point to which he has paid his fare, to quit such carriage, is for every such offense liable to a penalty of *forty shillings;* and any person committing such offense may be lawfully apprehended and detained by the company's officers and servants until he can be conveniently taken before some justice."— G. W. R. *of Eng.*

The laws of England protecting other companies are substantially the same as the above.

reason that the dropsical or apoplectic style of type is so much affected in railway literature.

The trains provided for in the schedule are called regular trains.

Each train has its number.

Trains going in one direction bear odd numbers, while those moving in a contrary direction monopolize the even numbers. Thus to hear the number of a train is to know its direction.

The relative importance of trains is indicated by the grade given them, as of the first, second or third order.

The number of grades may be restricted or indefinitely expanded.

The schedule fixes the grade of each train.

The life of a schedule varies from a day to six months.

It is the creature of circumstances.

The rules and regulations forming a part of the schedule accurately define the right possessed by each grade; sometimes of specific trains.

Thus the passenger trains northward bound are only required to wait five minutes at meeting points in the event trains of the same grade moving in an opposite direction are delayed, after that they proceed on their way, keeping five minutes behind their schedule time, until the belated trains are met.

But in the event a north bound passenger

train is delayed, the train going south is compelled to wait thirty minutes at the meeting point before proceeding; after that it resumes its journey, keeping, however, thirty minutes behind its time until it meets the delayed train.

Trains of an inferior grade are required to keep out of the way of those of a superior grade.

Thus if, at a meeting point of two trains of dissimilar grade the train of superior rank is late, the train of inferior grade must await its arrival indefinitely.[1]

If the case were the reverse of this, the train of inferior grade being behind time, the superior train would go forward without awaiting the arrival of the delayed train.

The regular trains provided for by the schedule are supplemented by others as business, or the exigencies of the service require.

These supplementary trains are known technically as extra or wild trains.

When one or more trains follow a regular train and are protected[2] by it then they possess

---

[1]. The number of hours a train must be behind time before it loses its rights as a regular train, varies with different roads from eight to twenty-four hours. After a certain time it is not recognized, and can only proceed under special orders, or in company with some other train.

[2]. Signaled. Two green flags by day, or two green lamps by night, carried on the front of an engine, indicate that an extra is following, possessing all the rights of the train carry-

all the rights of such regular train; in such cases they are termed extra trains.

If, however, a train is operated under special instructions, pursuing its way from point to point as ordered, without reference to the time indicated in the schedule for the movement of trains, then it is called a wild train.

The wild trains in motion upon a line are sometimes greatly in excess of the number of regular trains provided for by the schedule.

When the business of a road necessitates a temporary increase in the number of its trains, or when delay or accident overtakes those in motion, it is then that the telegraph is brought into use for the purpose of accelerating their movements.

**MOVEMENT OF TRAINS BY TELEGRAPHIC ORDER.**

It is exceedingly difficult, if not impossible, for a great number of trains of varying weight, advancing in opposite directions upon a single track road, to move with the regularity and precision necessary to enable them to meet and pass each other daily at the places designated in the schedule.

Many things conspire to accelerate or retard the progress of a train, such as the number and

ing the signals; upon some roads red is used instead of green, upon others white, upon others blue. In England the signal is carried on the *rear* of the train.

weight of its cars, the quality of its engine and the skill of its driver, the state of the track, the character of the grades, the direction of the wind, the density of the atmosphere, the activity of the station force, and the efficiency and industry of the conductor and his assistants. All these affect its movement. The train that moves forward without difficulty at the rate of fifteen miles an hour to-day, will barely be able to make ten to-morrow.

In moving trains by telegraph, all irregularities and inequalities are recognized and specifically provided for.

The trains are advanced from point to point without reference to the schedule.

The train that is running at the average rate of speed is moved ahead until the slower train is met, their meeting place depending altogether upon the exigencies of the hour, or their location as they approach each other.[1]

---

[1]. "The Superintendent arranges the schedule by which trains are moved, and when accidents occur, or business can be expedited thereby, the time table is superseded by the telegraph. To the discharge of this delicate duty he brings a clear head, attentive memory, and a perfect knowledge of the geography of his road, including the extent of its grades, the location of its telegraph offices, and the capacity of its sidings; the character, number and exact position of the trains in motion have accurately to be kept in mind, the quality of the engines hauling them, the state of the weather, the direction of the wind, and the peculiar capacity of the enginemen and conductors engaged."—*Railway Revenue, pp.* 34, 35.

When regular trains are moved by telegraph, they do not thereby lose the rights awarded them by the schedule, except so far as they may be specially affected by the orders they receive.

The moment a special order is fulfilled, or ceases to operate, the train it affects resumes the fixed rights it possesses as specified in the schedule; if a regular train, it conforms to that instrument; if a wild train, it awaits further instructions before proceeding, or seeks the protection of a regular train.

Special orders are rarely if ever issued that affect passenger trains, except when they are behind time, in which case the telegraph is brought into requisition for the purpose of expediting their movements, and at the same time keeping other trains in motion. With this exception, the orders issued may be said to relate exclusively to trains of inferior grade.

When there are a great number of freight trains in motion, in excess of those provided for by the schedule, or when they are for any reason delayed, they are moved by special order, without much, if any, reference to the time table.

In a central office the dispatcher watches the movements of trains and notes their wants. His is the master-spirit, and the various officials employed upon the road come and go as he directs without question or remonstrance.

Like the pieces on a gigantic chessboard, the

trains move in harmony with his will and are ultimately brought safely to their several destinations by him.

He constructs in his mind's eye a schedule adapted to the exigencies of the occasion. The requirements of this creation of his mind are known only to him.

He executes it with clearness, expedition and safety.[1]

Of course there are degrees of excellence in this field as in every other. The mind of one dispatcher will be clear, quick to apprehend and execute, the mind of another will be slow, heavy-witted, fatty. The movement of trains by telegraphic orders on a single track road requires an excellent memory and the exercise

---

1. "In the movement of trains much depends on the train-dispatcher, who fills a most responsible and laborious position. The latter-day train-dispatcher sits at headquarters, and, with the aid of a curious chart, is enabled to see at a glance the exact whereabouts of every train on the road at any minute of the day. He has the entire line before him in miniature. Dots and pegs of different size and shape indicate the different trains in motion at the same time, and from the chart and an elaborate time-card the train-dispatcher is enabled to direct operations by telegraph with as much intelligence and absolute knowledge as he cou'd possibly have were he ubiquitous, and able to give oral commands in a hundred different places at the same time. The train-dispatcher is supposed to know, and does know, the size of each train, freight and passenger, on his division, the speed and power of each engine, the grade of every mile of the road, and where time can be made up to the best advantage when trains are delayed."—*Newspaper account.*

of the nicest judgment at all times; where more than one track is employed the problems are greatly simplified, just as we have shown it to be less difficult to frame a schedule for roads possessing more than one track than it is where only one track is available.

The capacity of a single track road may be increased fully one hundred per centum, perhaps more, by a skillful use of the telegraph in connection with the movement of trains.

The statement appended hereto[1] of the performance of trains for fourteen consecutive days upon a single track road, 108 miles in length, with the usual station facilities and sidings, represents the perfection that has been attained in this important branch of railway service. The bulk of these trains were moved by the dispatcher through the medium of the

1. Total number passenger trains west bound, 56
" " " " east " 56
" " freight " west " 308
" " " " east " 301
——— 721
Freight cars in west bound trains, - - 7,701
Freight cars in east bound trains, - 7,272
——— 14,973
Average number of cars per train, - - - 24.59
" .. " trains per day of twenty-four hours, 51.50
" " minutes between trains at any point, 28
" distance run by trains, - - - - 94.70
" number of miles per hour, - - - 17.50

A still greater traffic could have been accommodated had the business of the line necessitated it.

telegraph; no accident or mishap of any kind attended their manipulation. The results indicate, of course, an alert and able dispatcher, and an efficient organization subordinate to him, but above all, they demonstrate the possibilities of a single track line when operated under favorable auspices.

### THE PROTECTION OF TRAINS.

A glance at the rules and regulations, including the signals, governing the movements of trains will convince the most skeptical of the careful forethought, the boundless provision made to ensure safety of life and property.

Wherever danger is to be apprehended, there signals are placed to convey to the far-off train assurance of safety, or warn it of impending disaster.

At night the lights of different colors that flash forth from the darkness as the train advances, guide the faithful engineman, just as the light-house on a dangerous coast serves as a guide and protection to the passing vessel. The daylight, however, affords the greatest latitude for arranging and displaying signals, and thus flags of varying color, strange symbols and quaint devices meet the gaze on every hand; these serve to stay the progress of the advancing train, or cheer it on its course.

The irregular, or working trains of a company, such as gravel, dirt, stone, and wood trains, constitute one of the greatest elements of danger.

This is especially so upon comparatively new lines.

The duties performed by these trains compel them to visit every part of the line at infrequent and indefinite periods, not hastening, like other trains, from station to station, but proceeding leisurely, stopping here and there upon the main track, as occasion requires, to load or unload. While those in charge of such trains are able, if discreet and watchful, to keep out of the way of trains operated in accordance with the schedule, they are unavoidably kept in ignorance many times of the number and location of wild trains in their vicinity, and thus those in charge of the latter are compelled to exercise the utmost vigilance to protect themselves from possible disaster. In cases of this kind they are required to advance slowly, sending a signalman ahead as they approach curves and obscure places in the track.

Those in charge of working trains are, as a rule, required to keep signalmen at least half a mile in each direction when the train is employed upon the main track, and when in motion they are, or ought to be, for obvious reasons, required to move at an exceedingly slow rate of speed, except when the view of the track is unobstructed for a long distance in advance.

Where trains move uniformly in one direction upon a track, the precautions necessary to protect them, in the event the road is obstructed from any cause, are very materially simplified, it being only necessary to guard the approaches from one direction. This is a matter of much greater importance than is apparent at first sight.

The obstructions to the track from delayed trains, from the replacing of rails, the repairs of bridges and culverts, and other changes and improvements, are of constant recurrence upon every line. These obstructions, that invite the destruction of advancing trains, must be carefully guarded by sentries placed far away in each direction, where only a single track is employed; but where two tracks are in use, signalmen are only necessary in one direction, and thus not only the expense is lessened, but a great, ever-present, possible danger is averted.

For it must not be forgotten that while it is possible to surround every contingency or incident of railway experience that may be said to be subject to the government of man with such carefully devised directions for the guidance of employés as to definitely insure the safety of trains, in the event the directions are faithfully observed, still no provision, no forethought upon the part of railway managers can avert the consequences of the indifference, the gross stupidity, or utter recklessness that we must sometimes expect where so many men are employed. It is of the utmost importance, therefore, that the possible contingencies that may arise requiring the exercise of the judgment of employés in cases of danger should be restricted as much as possible.

## CHAPTER II.

INDIVIDUALITY OF RAILROAD COMPANIES. DISSIMILARITY OF THE SIGNALS IN USE UPON DIFFERENT ROADS. THE DANGER THAT SUCH DISSIMILARITY RENDERS POSSIBLE.

The individuality that characterizes the organization of railroads finds many curious illustrations, but none more curious perhaps than the diversity that exists in the signals employed by them in connection with the movements of trains.

Now under all ordinary circumstances nothing is more to be commended in a railway company, perhaps, than strong, well-defined individuality. Individuality means advancement, better facilities, a higher ideal, and the company that does not possess it soon loses its progressive characteristics, becoming instead an absorbent, simply, a drone. But when this individuality is carried to the extent of enforcing a different set of train rules upon every line that may have a distinct management, then the skeptical and uninformed traveler begins to doubt its expediency or wisdom.

It is undoubtedly true that the safety of the

lives of passengers and others depends at times upon the intelligence with which signals are manipulated. Emergencies are not of rare occurrence where the employment of the right signal at the right moment, and the instantaneous interpretation of its true significance by the approaching train, has saved the lives of many people, and prevented the destruction of valuable property. Hence, it is apparent that the signals in use should be stripped of all unnecessary ambiguity, and reduced as much as is possibly consistent with a clear understanding of what is required under every possible emergency.

A correct understanding of the subject requires that we should remember that the men employed about our trains are not wedded to the services or customs of a particular line. They are cosmopolitan. The force employed upon a railroad is constantly changing; these changes are accelerated or retarded by various causes. A great increase in the business of a company, a strike among its employés, political disturbances along its line, sometimes render it necessary for a company to put untried men upon its engines and entrust its trains to strange conductors. These new men may understand generally the practical duties of their several places, but they are unacquainted with the peculiar signals and rules of their new employer. It is not unlikely that they have at various periods of their lives served upon many different lines. This varied service has familiarized them with the use of many different systems of signals, and herein lies the danger.

This confusion of knowledge may portend many things. To the reflective mind it never ceases to be a subject of prolific interest and speculation. This diversity of knowledge upon a subject requiring nothing but definiteness, singleness of purpose, arbitrary precision, possesses a sinister meaning—seems to be pregnant with disaster as certain as the coming day.

A mere looker-on perhaps over-estimates the importance of the signals that meet his eye in every direction as he is whirled through town and village at dead of night. He has, perhaps, in his time, passed through some great accident, and its horrors have made him timid. Such people are very observing.

He has remarked that upon the Great Blank Road a green light is a signal of caution, a signal to trains to moderate their speed; it does not tell them to halt. Upon the Great Trans-Continental Line green is a signal of danger; its warning is imperative, absolute; it says, *Stop that train!* not at some indefinite point beyond, but there; there where the lamp burns; not a foot further—death lies beyond.

But suppose the engine-driver has but recently come into the employ of the Great Trans-Continental Company after many years of faithful service with the Great Blank Line; enginemen are always making changes of this description. A skillful mechanic and noted for his watchfulness and fidelity to duty, he is a valuable acquisition to the road and, after a month or so, he is put upon the night express. This train is always heavily loaded; it makes no stops, and keeps pace with the flying clouds. As it plunges forward through the darkness the engineer observes everything, and as he rounds a sharp curve a green light, shining upon the track before him, meets his gaze; he has seen it under similar circumstances many times before; its reflection gladdens his heart like the face of an old friend; it relieves the monotony of the dark night; he approaches it cautiously; such has been his custom. The green lamp is to him like the warning of a comrade when one glass more might take him off his feet; it is a good-natured nod to go slow; it is not imperative. As the train rolls by he leans lazily out of his window, but the signalman, wild with rage and fright, hurls the lamp full at the cab, and it is smashed into a thousand pieces. In an instant the truth flashes upon the driver; upon this line green is a signal of danger; a chill of horror seizes him; he is running at the rate of fifteen miles an hour; he reverses his engine, the whistle sounds, the brakes are screwed down, the drivers whirl in reverse circles with the velocity of light, the engine sways and trembles with the tremendous strain put upon it, but it is too late, and there far away in the country, in the peaceful stillness of night, the great black engine, its brave driver and long line of cars filled with sleeping passengers, glide quietly, imperceptibly, into the yawning gulf that envelopes them all in a common ruin.

This is what a diversity of signals means to the tired and nervous traveler.

How far are his fears justifiable?

Could the case we have supposed actually occur?

Probably not.

Yet it is true that the signal that correctly interpreted says to the engine-driver, "all right; go ahead; the track is clear;" may, and undoubtedly does, mean something entirely different upon a neighboring line.

Accidents occur upon our railways that are inexplicable.

The occasion of them is enveloped in mystery.

The religious attribute them to God.

Are any of these disasters brought about by an improper understanding of the meaning of particular signals, or by employés getting the rules and regulations of different companies confounded? Who can tell?

An investigation of the subject of train regulations elicits many curious things. Upon one great line the carrying of two green lights in front of an engine is a notice to the trains it meets that the track is clear; no trains are following; go ahead. Upon another great line two green lights carried upon an engine indicates that a train is following and that all other trains must keep out of the way. These signals mean two directly opposite things, and a conductor and engineer, acting upon the signals of the first mentioned company while in the employ of the second, would inevitably bring his train into collision with another, if no fortuitous circumstances intervened to prevent it.

The lamp raised and lowered upon one road says back up; upon a parallel line, not ten feet away, it may, and very likely does, mean, go ahead.

Differences like these are pregnant with ideas of danger.

### WANT OF UNIFORMITY IN THE RULES AND REGULATIONS GOVERNING DIFFERENT ROADS.

It would naturally be supposed that where a track was used in common by two companies, that their system of signals, and their rules and regulations, would be identical, not only upon the track that was jointly occupied, but over all of their lines as well. To a superficial observer the danger of getting the two confounded would seem to be so great that their unification would follow as a matter of course. But it does not. It is impossible to tell why. It may not be necessary. We will believe that it is not.

But a single section of track can not safely be operated under conflicting rules, and when it is used by two companies one of them must necessarily give way. Obviously the company to give way will not be the proprietors of the track. Accordingly the other company will direct its employés to observe the rules of the proprietors of the joint track when passing over any portion of such track. These employés must, therefore, at a particular place, be it day or night, lay aside the rules and regulations that they are familiar with by study and practical use, and adopt in their stead other rules dissimilar in form and application.

It would seem as if it could not be otherwise than hazardous to make this abrupt substitution of rules under which trains are operated day after day, and year after year. But we must believe that it is not, for there are instances where a joint track has been operated for a series of years under the very circumstances we have mentioned.

### INTELLIGENT DISCRIMINATION EXERCISED BY TRAINMEN.

It is observable in the practical application of the system under which trains are operated, that the employés connected with the train service do not always attach the significance to specific signals or rules that would naturally be supposed. Especially is this so in reference to use of signals. Their acquaintance with the every-day working of trains teaches them that allowance must always be made for the ignorance, stupidity or thoughtlessness of employés, and they strive constantly to protect themselves and the passengers and property entrusted to their care from the fatal effects that would oftentimes follow a blind obedience to the orders given them by the class of men we have enumerated.

And so it is in reference to special orders. The engineer of an irregular train that is running under special telegraphic instructions at the rate of sixty miles an hour, can not depend implicitly upon the accuracy of the reports he receives in reference to the location and intention of other trains. Doubtless the information imparted to him is perfectly accurate and trustworthy. He ventures no comments. His orders are to proceed. He has been trained to obey.  Outwardly, he is unconcerned, but inwardly he is filled with apprehension, and as he proceeds on his course, he scrutinizes the track with an intensity and a sagacity that never wearies.

The anxiety upon the part of the engineer is not occasioned by fear for his personal safety, though that doubtless has its influence, but it is the knowledge, born of observation and experience, that blind adherence to orders, no matter what the circumstances or from whom emanating, may not only cost him his life, but may involve the lives of many others; the lives of people believing in him, and trusting him, and as unconscious of danger as they are helpless to avoid it.

Under many circumstances the watchfulness of the engineer is of no practical avail; a sharp curve may bring him face to face with an advancing train, an open switch or a track torn up for repairs.

Some rule upon which his safety depends is disregarded. The train that should wait proceeds on its way confident of making the succeeding station; the night is foggy, a high wind blows, the track is slippery, the engine will not make steam, its time is up. Still it advances; when from out the gloom there emerges in its immediate front the light of an approaching locomotive; the whistles simultaneously shriek the alarm; there is a moment's suspense; when high above the roar of the winds, and the noise of rushing steam, is heard the crash of the opposing trains.

### THE CONSERVATISM OF TRAINMEN.

That disasters of this character are of rare occurrence is attributable to the intelligence and watchfulness of the men in charge of our trains.

A disregard of the established rules under which trains are manipulated, not only costs the offender his place, but it may involve many innocent lives.

This tremendous responsibility can not be evaded, and so there grows up in the mind of the engineer and conductor an intense conservatism.

Subordinate employés participate in this feeling, and so we find everywhere we go a disposition, upon the part of trainmen, to comply with the literal requirements of each and every order or rule, and in cases of doubt nothing is risked, everything is sacrificed that absolute safety may be ensured; and it is to this conservatism, this loyal adherence to established rules, that the railway traveler is indebted for his safety.

## THE REGULATIONS PARTAKE OF THE CHARACTER OF THE MEN INTRODUCING THEM.

As we advance in our inquiries into the rules governing the machinery of the department of transportation upon different roads, we are more and more surprised at the differences that exist.

Many of the differences are material.

Others, again, are differences of form, only.

In many cases we can trace in the regulations of a road the peculiar traits of character possessed by those instrumental in perfecting them.

The rules of one company will be extremely exacting; another company will trust more to the discretion of its operatives.

Much can be said in favor of each system. Under one system employés act automatically; under the other they act more zealously, perhaps, but with less effectiveness. The first named system is without doubt best for the company, the last named is more advantageous to the men. Generally speaking, one system breeds dependents, the other engenders men.

### SOME OF THE DIFFERENCES OBSERVABLE IN THE RULES AND REGULATIONS OF DIFFERENT ROADS.

But let us notice further some of the differences that exist in the regulations of different railroads.

And first we remark that upon one line the trains going south possess certain privileges over trains going north; that is to say, they are entitled to the road for a certain specified number of minutes over and above the time allotted them in the time-table, and connecting trains are required to keep out of their way. Upon a neighboring road the trains going north will be the ones that are favored.

It does not require a vivid imagination to picture the consequences of any mistake as to the rights possessed by a particular train, but as a mistake in this respect must involve a misapprehension of the facts upon the part of both the engineer and conductor, it may be said to be improbable if not impossible.

The direction in which the greatest average number of people travel varies in different sections. In one section it will run to the north, elsewhere the stream will be southward. The discrimination we have mentioned is usually in favor of that current of travel that it is most important the railway company should favor.

The granting of certain privileges to a train moving in one direction, not granted to trains moving in an opposite direction, is, therefore, not the result of chance or caprice, but the exercise of a shrewd discretion.

In pursuing our investigations, we find constant evidence of the exercise of this discretion. One company will insist upon its gravel and other working trains keeping ten minutes or more out of the way of all goods trains, that is to say, they must be clear of the main track at least ten minutes before a freight train is due. These working trains employ hundreds of men, and in the event the freight train is delayed, or whether it is or not, the loss of money to the company through the enforced idleness of its men must, in the course of a year, amount to a large sum. A neighboring company, keeping this fact in mind, will give its gravel trains permission to continue at work (keeping out the required signals) until the approaching freight train is in sight, when the working train must hasten to get out of its way. Under this rule no time is lost unnecessarily by the employés of the company, and under its practical working it may be entirely safe, though examined theoretically it would seem as if the order requiring working trains to keep at least ten minutes out of the way can not but be safer than the rule permitting them to continue at work, no matter what careful provision may be made for watching the approaches to such train.

The margin of time allowed trains of a superior class, which time must never, under any circumstances, be used by those of an inferior order, is not the same upon different roads. One company will require its freight trains to be upon a siding twenty minutes in advance of the time a train of superior grade is due. Upon another line fifteen minutes will be allowed. Upon still another road ten minutes is considered sufficient. The object of each management is, of course, to strike a happy mean. The safety of trains, and especially those of a high grade, is always of paramount consideration, but a due regard for their safety is not necessarily inconsistent with an active, expeditious discharge of business, and if a margin of ten minutes is considered sufficient by the management, and has been proven to be so by years of experience, then to allow a longer time would be an unnecessary delay of the traffic of the line, and a gross extravagance upon the part of the company's representatives.

An effort upon the part of railway managers to make the most of every circumstance is apparent in many ways. That these efforts at economy are often times illy directed and unfortunate in their results is made apparent from time to time, and these failures teach us to remain silent when we would otherwise be disposed to criticise what seems like a want of thrift, an improvident use of the resources of a road.

One company will require its detached engines, when passing over the line, to precede, in all cases, the regular trains. Another company, with a careful eye to the saving of a few pence, will require that when such engines accompany freight trains they must follow rather than precede. The object of the latter case being, doubtless, to make the detached locomotive assist the engine attached to the train in the event assistance is required. The danger to the train and its operatives is apparently much greater from an engine following, than from an engine preceding it, but the opportunity of using the detached locomotive, as occasion requires, is thought to more than compensate for the slight risk that is run. Whether it does or not no one can definitely determine.

Of the many differences that attract our attention, not the least surprising is that which exists in reference to the manner of conducting business upon double track roads. While it seems perfectly apparent to us that vehicles should, to prevent collision, turn to the left in passing each other upon the public highway, it also seems equally clear that upon a railway line where the danger of collision does not and can not exist, that trains should in all cases take the right hand track. As the regulations of the English companies and many of our own lines require that trains shall run upon the left hand track, we must accept such regulations as conclusive evidence, that in the estimation of the managers of such lines at least, there are many weighty reasons why trains should run upon the left hand track in preference to the right.

### NO UNIFORMITY IN THE TELEGRAPH DEPARTMENT.

The diversity that exists in the rules of different companies governing the movement of train operatives also exists in the telegraph department of railroads.

Upon the lines of one company, the signal "27" flying along the wire closes every key and silences every operator; it is a magic number; it hushes all disputes; it means life and death; it is a warning to clear the line; it is a signal that the waiting message must take precedence of every thing else, no matter how important.

Upon another circuit "27" possesses no significance whatever, and its repetition would never still the struggle that is forever going on amongst operatives for the use of the line.

Upon one line the cabalistic sign "19" serves instantly to hush all rivalry and contention, it is the signal of the general manager, and woe to the unfortunate novice who incautiously ventures to break in upon the business that follows. Upon another line number "19" has no special meaning, and its repetition would only serve to excite idle curiosity or profanity.

Upon some of the telegraph lines the most extended and ingenious ways are sought to abbreviate and save time. Each number will be made to convey some special information, an elaborate question perhaps, while still other numbers furnish an answer for every emergency. When this field has been exhausted, the alphabet will be resorted to and isolated letters or simple combinations of letters will be made to stand for words, the words selected being those most in use in the business vocabulary of a railroad.

## LACK OF COMPLETENESS AND THOROUGHNESS IN FRAMING THE RULES AND REGULATIONS.

While we find that the rules of no two companies are exactly alike, so we find on a careful examination of the regulations of many different lines that no one of them contain all the rules that possess a positive practical value ; no one of them that is not deficient in some important respect.

Investigation elicits the fact that the rules of each company contain many valuable hints and suggestions not embraced in the directions of any other company.

It has been the aim to embrace in the rules appended hereto the salient features of each, so far as the same was practicable. The magnitude of the work has for the present rendered the effort only measurably successful.

The bits of information gleaned in pursuing the wise provision made by different managers are interesting as well as instructive. One manager who has, doubtless, in his time given special attention to the subject of claims, directs his subordinates in all cases of accident to report with other facts the names of witnesses. He has undoubtedly been sorely pressed by some unfriendly claimant in consequence of lack of information upon this very point. Other companies note the provision made and insert similar instructions.

The same manager we have mentioned also warns his employés in his printed rules that his company will not under any circumstances be responsible for accidents to employés while coupling cars, etc. Evidently he does not intend his company shall suffer from negligence in this particular field, if warnings will suffice.

Another manager will take a rule common to all roads and, by adding a clause, perhaps a word, give it a finish and completeness that it did not before possess.

Another manager explains to the operatives of his trains that they must not exceed fifteen miles an hour, and that, when running at that rate, they will pass seven telegraph poles a minute. Probably this would be only approximately true upon many lines. It is however a fine illustration of the acute observation and good practical sense of railway managers.

Another manager provides a system whereby trainmen may signal each other in the event a train should break in two, special provision being made in case the train should break into more than two parts. We should probably find upon inquiry that the company represented by the official promulgating these signals, has at some period of its existence suffered disastrously from the inability of trainmen to convey quick intelligence to their companions of the breaking in two of trains.

Still another manager is at considerable pains to define the rights possessed by an extra train, in the absence of special orders, in the event it can not reach the meeting point without trespassing upon the time of trains going in the opposite direction.

And so we might go on at much greater

length, but enough has been written to illustrate the differences that exist in the laws governing the movement of trains, and here for the present we drop the subject.

**THE WONDERFUL PHRASEOLOGY OF TRAINMEN.**

Of the many remarkable things noticeable in the experience of railroads, not the least curious are the technical phrases in common use, in connection with the train service. Many of the phrases are, without doubt, re-adaptations of old expressions common to the early experiences of the pioneer managers of railways. The necessities of the service have given rise to many other expressions peculiar to it, and not to be found elsewhere.

While the words and set phrases employed are perhaps not as copious or extended as those in use among sailors, still, many of them are quite as enigmatical, and to any one ignorant of their application they possess a significance that is startling in the extreme.

Thus, when the ukase of the manager goes forth that "flying switches" will not be tolerated upon the line under any conceivable circumstances, the verdant observer is quite justified in picturing in his mind's eye an ingenious contrivance whereby certain vicious and unruly employés are accustomed to amuse themselves, surreptitiously perhaps, from time to time, to the great distress and alarm of the management.

Our verdant friend finds that "running switches," "shooting stations," and "wild trains" are everywhere spoken of as the most natural and proper objects in the world — things too well known and understood to require elaboration or explanation.

And in this way his mind becomes expanded, so that when he reads that "enginemen must not fail to note all 'whistling-posts' they may pass upon the line," he is neither daunted nor discouraged, but at once acknowledges and accepts the presence of "whistling-posts" as he would any other phenomena in nature.

However, when he reads that conductors will "side-track," under certain stated circumstances, he is at a loss to know whether their doing so will be voluntary or involuntary. Are they to side-track of their own accord, or will they side-track in spite of themselves?

These and similar questions constantly recur to disturb him as he progresses; they can not be absorbed, and are too enigmatical to be solved unaided.

When he reads the terse command that conductors must "take a side-track," he wonders, inwardly, if they take it as they do medicine or food, or, as an outlying fortress is taken by storm with attendant sappers and miners.

And so he wonders how it is possible to turn trains upon the letter "Y," and why so foolish a thing should be done.

He can not understand why it should be necessary to tell a man of sufficient intelligence to act as conductor, that he must "keep off" the time of other conductors, and speculates what connection, if any, this has with the "lost time" of trains.

What process is necessary to enable one train to "clear" another?

Why should not an engine be allowed to slip her "drivers" if she or they can get along easier thereby?

How are switches "set," and in what manner can a train be operated upon a "block?"

Questions like these occur to him at every step.

In another chapter we have endeavored to explain the meaning of some of the more obtuse phrases common amongst trainmen. Some of these phrases are well understood, others again are unintelligible, except to those versed in what we may call the phraseology of trains. The list is susceptible of infinite expansion, but it is sufficient in its restricted form for the purposes of the present work.

## PHRASEOLOGY PECULIAR TO ENGLISH ROADS.

While the phraseology employed upon English roads is radically different from that in use in this country, it is in no respect less peculiar. Yet it is probably true that any Englishman who should attempt to explain the phrases in common use upon the roads in Great Britain would be generally laughed at by railway men in that country; to them such phrases are a part of their mother tongue; by many they are supposed to be in universal use; by others they are thought to have always formed a part of the English language. Yet, while the English language is still tolerably well understood in the United States, it is nevertheless true that there are probably not one hundred Americans connected with the various railway companies in this country who understand the significance of the great bulk of expressions in common use upon the railways of England.

How many Americans are there who know what a Scotch block[1] or a sprag is; or a trolley,[2] lay bye,[3] lorry,[4] ganger,[5] or train staff[6]? This list could be extended indefinitely.

1. A block laid across the track to prevent the movement of cars.
2. Car used by trackmen.
3. A side track.
4. A flat car.
5. The foreman in charge of sectionmen.
6. A staff used upon a single track road and placed in a

In England, as in the United States, the names of many things connected with the railway lines had a significance half a century ago that they do not possess under the new order of things.[7]

socket upon the engine to indicate that such engine has been granted the right to run over a particular section of line.

7. "At the 'booking-office' no booking is done. You merely say, to an unseen if not invisible person, through a small hole, 'First (or second) class, single (or return)' put down your money, receive your ticket, and depart. But as there were booking-offices for the stage-coaches which used to run between all the towns and through nearly all of the villages of England, the term had become fixed in the minds and upon the lips of this nation of travelers. So it was with the guard and his name; and when the railway-carriage supplanted, or rather drove out, the stage-coach, the old names were given to the new things, and the continuity of life was not completely broken."—*Richard Grant White in the Atlantic.*

## CHAPTER III.

EXPLANATION OF SOME OF THE TECHNICAL TERMS IN USE IN CONNECTION WITH THE TRAIN AND STATION SERVICE OF A RAILWAY COMPANY.

*Ahead of Time.*—When a train reaches a place before it is due at such place, according to the schedule or special order under which it is running, it is said to be ahead of time; in advance of its time.

*Behind Time.*—When a train fails to reach a point at the time specified in the schedule or special order under which it is operated, it is said to be behind time; when a train is late.

*Block System.*—A system devised for the expeditious movement of trains without jeopardizing life or property. Under the block system the track of a road is cut up into short sections of a few miles in length called blocks. Not more than one train is allowed on a block at a time, except as noted below. When a train passes off from a block the fact is at once telegraphed to the operator at the opposite end of such block; the track thus becomes free for the use of any following train. Until receipt of this notice no train is permitted to enter the block without specific notice in each instance that the block is already occupied and that its speed must be governed accordingly. Under the block system the officials of a train are warned and the train is itself protected when the road is obstructed by preceding trains.

*Brake.* — In railway parlance an apparatus attached to engines and cars for the purpose of bringing them under more complete control, to be used when occasion requires in lessening their speed or stopping them when in motion.

" A piece of mechanism for retarding or stopping motion by friction, as of a carriage or railway car, by the pressure of rubbers against the wheels." — *Webster.*

The application of this power or friction to the wheels is called " setting the brakes," "set the brakes," " the brakes are set."

*Cars.*—The cars employed by a railroad in the conduct of its business may be enumerated as follows, viz: In passenger trains, baggage, business, directors, drawing - room, express, hotel, mail, milk, officers, palace, passenger[1]

---

1. Passenger cars are called coaches or carriages in England. In Europe the passenger cars are divided into compartments, with separate entrances on each side of the car. The compartments of first-class carriages usually contain seats for eight, four on each side. In the lower classes there is no partition between the seats. and a greater number of passengers can consequently be accommodated. Passengers in different

first-class, passenger second-class, parlor, pay, saloon, sleeping, and smoking. In freight trains, boarding, box, caboose, ditching, dump, flat, freight,[1] horse-boxes, mineral, oil, ore, paint, pile-driver, platform, refrigerator, stock, way, and wrecking.

*Classes of Trains.*—" Regular," " Extra," and " Wild."

*Clearing a Train.*—Keeping out of the way of a train. Arriving at a meeting, or passing point, before the train to be cleared is due. As " clearing a train ten minutes."

*Closed Switch.*—When a switch is " closed " the principal, or main track, is uninterrupted, continuous, not diverted.

*Construction Train.*—A train employed exclusively in the transportation of material belonging to, and used by, a railroad company in connection with the improvement of its property, or the building of new lines. It usually embraces trains engaged in hauling ballast, dirt, gravel, stone and timber, or employed in removing earth from ditches and cuts. Trains occupied in the work last described are frequently called ditching trains.

compartments of first and second-class cars can not communicate with each other (the partitions extending to the ceiling) and are isolated from the officials in charge of the train. The water-closet to be found in all of our passenger cars is unknown abroad.

1. Called wagons in Great Britain.

*Extra Train.*—A train not expressly contemplated or provided for in the schedule. It is run for the purpose of expediting the business of the road; to accommodate the traffic that can not be hauled in the regular trains without delay. It follows a regular train usually of its own grade and possesses the same schedule rights as the train it is following.

*Flying Switch.*—The disconnecting of a portion of a train while in motion and just before reaching a switch, the forward part of such disconnected train accelerating its speed to such a degree as to enable it to reach and pass the switch in time for the person in charge thereof to divert the detached cars that are following, to some other track.

*Grade of Trains.*—The grade of trains varies upon different roads, but it may be stated, approximately, in order as follows: The first grade embraces the four classes of passenger trains, viz: express and through mail, local mail, suburban, and accommodation. The second grade embraces the three classes of freight trains, viz: live-stock, through, and way. The third grade embraces the wild trains, viz: the trains operated under special or telegraphic orders, including construction and wood trains.

*Holding a Train.*—Delaying a train for any reason. A train may be held for orders; until some other train arrives; until a brake can be repaired.

*Irregular Train.*—See "wild trains."

*Keep off the Time of a Train.*—A direction not to obstruct the main track or attempt to occupy it when, according to the schedule, it rightfully belongs to another train.

*Lost its Rights.*—See "when a train has lost its rights."

*Lost Time.*—The time that a train has lost, taking the schedule as a basis. If a train is behind time it may be said to have "lost time."

*Main Track.*—The main track or tracks of a road upon which its trains are run.

*Making Time.*—Signifies that a train is running in accordance with the time allotted it in the schedule; is not losing time.

*Meeting Point.*—A point at which trains moving in opposite directions meet.

*Movement of Trains by Telegraph.*—Telegraphic orders directing the movement of trains. The manipulation of trains from a central office through the medium of orders sent by telegraph. The substitution of special orders for the fixed time and rights allotted trains in the schedule, and in the rules and regulations appertaining thereto. Directing what trains shall have the right to the road, and where and when they shall run without reference to the rights allotted them in the schedule.

*On Time.*—Means that a train is conforming exactly to the time specified in the schedule; in accord with it.

*Open Switch.*—When a switch is "open" the main track from one direction is connected with a subsidiary or collateral track, while the main track from the opposite direction is not connected with anything. " Open a switch " is to disconnect the principal track and connect one part of it with some other track.

*Overshooting.*—Running past a point as "overshooting" a station.

*Passing Point.*—A place where a train is overtaken and passed by another train going in the same direction.

*Regular Train.*—A train specifically named and graded in the schedule, as "Passenger train No. 3."

*Right to the Road.*—The right of a train to proceed on its course. The right to occupy the main track at a particular time and place, to the exclusion of all other trains of the same or inferior grade. In the absence of special orders to the contrary, trains of an inferior grade are required to keep out of the way of trains of a superior grade, *i. e.* when a train of a superior grade is due according to the schedule, trains of an inferior grade must not occupy the main track until the superior train has passed.

*Rights of a Train.*—Certain rights that a train possesses as defined by the schedule and the rules and regulations governing the movement of trains. The right a train has to proceed according to the time allotted it in the schedule, when it can do so without impeding the course of a train of a superior grade, or when not otherwise ordered. The rights a train of a superior grade possesses over trains of an inferior grade. The rights under certain circumstances which a train going in one direction possesses over trains going in the opposite direction, etc.

*Running Against a Train.*—When two trains are to meet at a certain point they are said to be running against each other.

*Running Time of Trains.*—See "Time."

*Run Regardless.*—A special or telegraphic order to run a train regardless of another specified train or trains. An order giving a train the right to the road against another train as " You will run from Fort Edward to Glens Falls regardless of train No. 9, but keeping out of the way of all other regular trains."

*Schedule or Time Table.*—The schedule accurately fixes the grade of each and every regular train; it provides where trains shall meet or pass each other; it fixes the maximum speed of trains, and gives each regular train a definite number, and specifies the time of its arrival at and departure from stations.[1] The rules and regulations governing the movement of trains

---

1. The schedules published in the various railway guides are in form substantially the same as those used by trainmen.

properly form a part of the schedule, and with these it is the chart that in the absence of special or telegraphic orders to the contrary governs the movements of trains.

*Semaphore.*—" An apparatus or piece of mechanism for exhibiting signals to convey information from a distance."—*Webster.*

*Setting a Switch.*—Arranging a switch so as to connect certain specified tracks. When a switch is adjusted so as not to disconnect the main stem, it is said to be "set for the main track." The directions to trainmen and others, so often to be met with, to see that "switches are set right," means that they are to see that switches are so adjusted as not to disconnect the main track.

*Shunting.*—The English term for switching.

*Side Track.*—A track varying in length and running parallel with the main track, and connected with it at each end by a switch. With unimportant exceptions, the freight cars required to transport the traffic of railroads are loaded and unloaded while standing upon these tracks; the tracks at the different stations vary in number and length with the business that requires accommodation. For the purpose of enabling trains to meet and pass each other upon the road, side tracks of varying length are required to be located at convenient points along the line. The terms familiar to railway men, "will take a side track," "will side track," means, when robbed of the peculiar phraseology in which they have been clothed, that the train referred to must run upon and occupy a side track.

*Sidings.*—See "Side Track."

*Signals.*—Train signals. The medium by which under certain circumstances intelligence is conveyed quickly, and at a distance between employés at night and by day, through the medium of the human senses. The signals consist of motions of the arms and body; of explosives or torpedoes placed upon the track; of flags and other devices of different colors for use during the day; of lamps of varying color and significance for use at night, and, finally, of information conveyed through the medium of the semaphore. Certain letters, figures, and combinations are in common use as signals upon telegraph lines for the purpose of expediting business.

*Slipping the Wheels.*[1]—When the wheels do not revolve (the engine or train being in motion) they are said to slip.

*Special Train.*—A train provided for a special purpose. It is not named in the schedule, and is moved under the special orders of the Superintendent. A wild train.

Trains of a certain character or grade, like

---

1. It is termed "Skidding the Wheels" in Great Britain.

suburban or way-passenger trains, are designated as special trains upon some lines. Upon still other roads, what we have already classified as an extra train, is called a special train.

A special train is an extra train in this, that it is operated for the purpose of meeting a want that the regular trains do not adequately provide for.

*Spur Track.*—A track connected at one end with the main track; it sometimes runs parallel with the latter, the same as a side track. These tracks are constructed for the purpose of giving a company access to gravel pits, stone quarries, and outlying manufactories and business enterprises, etc.

*Station.*—A place where the passenger traffic of a railroad, and much of its freight traffic as well, is received and discharged; the depot and its immediate vicinity. In the movement of trains a side track located at an isolated point on the line, possesses, in many important respects, the same significance as a station; a place where trains meet or pass each other.

*Switch.*—A mechanical apparatus constructed at the junction of two or more tracks, or at points where one or more lines diverge from the principal track. It is operated by a lever and cross bar, and by its aid lines diverging from the principal track are connected or disconnected at pleasure with the latter.

"To turn from one railway track to another."—*Webster.*

*Switching*—*Sometimes called "Shunting."*—The transfer of a car from one track to another. The manipulation of cars in yards and elsewhere. The arranging and rearranging of cars in making up trains so as to get them in the order desired. The arranging of cars upon the arrival of trains at their destination or while *en route.*

*Third Track.*—A third track or siding placed between the main tracks of a double track road for the purpose of enabling trains to pass each other with facility and dispatch. A track occupied by trains of an inferior grade for the purpose of allowing trains of a superior grade to pass.

*Through Train.*—A train designed to accommodate the through traffic or (in the case of a passenger train) the traffic between the large cities at which it stops.

*Time.*—The time allotted to trains by the schedule and by which their movements are governed. In some cases, though rarely, special orders are given to trains to run to a specified point in the event they can reach such point by or before a certain time named in the order.

*Time Table.*—See "Schedule."

*Train Dispatcher.*—An assistant of the

*Superintendent.* The official who directs the movement of trains by telegraph; an expert.

*Trains.*—The trains operated upon our various railroads may be specified as follows, viz: ballast, coal, dirt, excursion, freight, gravel, mineral, oil, ore, passenger, pay, stock, stone, timber, wood, and wrecking.[1] What are called "freight trains" may be said to embrace practically all the trains engaged exclusively in transporting merchandise and other property for which a railway company receives pay.

*Turn a Switch.*—To "turn a switch" is to disconnect one track from the main stem, substituting another track in its place.

*Turn Out.*—See "Side Track."

*Way Bill.*—An itemized account of property transported; a statement of the articles, the amount of the charges, the point from and to, date, number, etc.

*Way Train.*—A train that stops at the various stations and is occupied in doing the petty or local business of a company. An accommodation train. A way passenger train or way freight train stops at all regular stations. The duties of employés on way freight trains are multifarious as well as arduous. In addition to the ordinary duties of trainmen they are compelled to handle much of the freight hauled in

---

1. In England a wrecking train is called a break down van train.

their trains. For instance, a freight car sometimes contains freight in small quantities for several different points. It is the duty of trainmen to unload this freight. When the freight to be shipped from a station is not sufficient to warrant the exclusive use of a car, it is piled upon the depot platform to be loaded by the trainmen into some empty or partially loaded car. The engines of way freight trains do the switching required at the small stations.

*When a Train has Lost its Rights.*—A regular train, when twelve hours behind time, loses its right to the road against all regular trains. It is no longer recognized or provided for by the schedule. It ceases to be a regular train, and it is classed thereafter as an extra or wild train.

A train may lose its right as against a particular train or trains, and still possess rights that are paramount over those of other trains. Upon a single track road a train of the highest grade going in a certain direction is not allowed to leave a station where it should meet another train of its own grade, until thirty minutes after its leaving time. Thereafter it proceeds on its course, keeping thirty minutes behind its time, and the opposing train must keep out of its way.

Trains of an inferior grade cannot proceed until trains of a superior grade that are due, or past due, have arrived, unless the latter are twelve or more hours behind time.

*Whistling Post.*—A post or board erected in the vicinity of stations and crossings. A signal to the engineman to sound the whistle of his engine.

*Wild Train.*—An irregular train for which no provision is made in the schedule. It is operated under orders from the Superintendent, and is required to keep out of the way of regular and extra trains.

*Wood Train.*—A train engaged in hauling the wood required by a railway company for its own use.

**Y.**—A track of the general shape of the letter Y. A track connecting two tracks running at right angles with each other. This track, or combination of tracks, affords a convenient means of turning trains or cars.

## CHAPTER IV.

PLAN PURSUED IN ARRANGING AND COMPILING THE RULES AND REGULATIONS.

The accompanying directions in reference to train and station service have been compiled without prejudice from the rules and regulations in force to-day upon some twenty of the greatest, most thoroughly organized, and best managed roads upon this continent. The workings of all the principal roads of Great Britain have also been studied, and such of their rules and regulations as were thought applicable to our system of management have been embodied herein. In many cases where their regulations were not directly or wholly applicable, they have nevertheless been inserted as foot-notes for the purpose of illustrating their theory and its peculiarities, and for the valuable information and instruction they afford.

In compiling these instructions, it has been necessary in many instances to decide between conflicting rules. In such cases preference has been given to those that seemed under all the circumstances of the case to be the most feasible, or that possessed the greatest practical value in the economy of railway management.

The object of the compiler has been to form from the regulations now in force upon various lines a more perfect code of rules. It is doubtless true that this object has only been partially attained.

The compilation has not been made with the view or expectation of its adoption by any particular company. However, wherever the rules are applicable or valuable to railway managers, either wholly or in part, they will in time undoubtedly be accepted; where they are not applicable, or best, they ought not to be adopted, and will not be.

While it has been the aim of the writer to make the regulations embodied herein practicable upon any of our lines, it is nevertheless true that many rules that are imperative upon one line possess no relevancy elsewhere, or, more properly speaking, they are, under ordinary circumstances, unnecessary. The double track road, for instance, does not require rules so elaborate as those governing the use of a single track, still it is necessary to provide rules sufficiently comprehensive so that in the event any accident restricts a company to the use of one track the safety of trains will not be endangered nor the business of the road impeded.

The company that can, without inconveniencing the public, allow twenty minutes between its trains, will possess rules that, while they are wise in their application by that particular company, would be cumbersome and impracticable upon a line where the business required that trains should arrive and depart every five minutes, as is the case upon certain English roads during particular portions of the day.

The main purpose of the compiler in preparing these instructions has been to place within the reach of railway men, of every grade and occupation, facilities for acquiring accurate knowledge of the extent and scope of the duties and responsibilities of train and station men under the system of manipulating trains generally prevalent in the United States.

An examination of the rules and regulations of the best managed companies makes it apparent that many seemingly trivial but really important things that employés should possess accurate knowledge of are no where mentioned; it being accepted as a matter of course that the employés possess the desired knowledge. And it is doubtless true that those familiar by long experience with the practical working of trains do possess this knowledge, but the novice or student finds the omissions of a character not to be overcome except by long experience or diligent and protracted inquiry, which but few of them are able to prosecute successfully. The writer has therefore introduced new rules and explanations wherever he believed they would tend to a clearer understanding of the subject. And in reference to the construction of the old rules adopted by him, he has not hesitated to alter or amend their purport or phraseology wherever he believed greater efficiency or clearness could be secured by such alteration or amendment; the object being so far as possible to frame a code of rules sufficiently comprehensive to cover great enterprises as well as comparatively unimportant or partially completed ones.[1]

[1] The more minute rules and regulations of the block system having no general significance in the United States, and not being likely to have for many years to come, have not been embodied herein.

## CHAPTER V.

SIGNALS REQUIRED BY RAILWAY COMPANIES.

Flags of the proper color must be used as signals by day, and lamps of the proper color must be used at night or in foggy weather.

"Signal lamps must be lighted as soon as it commences to be dusk, and, during the interval between daylight and dark, both day and night signals must be used."[1]

Hand-lamps and hand-flags, when used as signals, must always be held in the hand, and not placed upon, or stuck into, the ground.

Red signifies danger, and is a signal to stop. It must never be used as a caution signal.

Green signifies caution, and is a signal to go slowly.[2]

In the absence of a green light, a white light waved slowly, from side to side, must be used; it denotes danger,—Stop.

1. English Clearing House Standard.
2. Out of fifteen American roads examined, eight of them do not use green as a signal. Upon one road it indicates, when carried upon engines, that another engine is following, and that such engine possesses all the rights of the engine carrying the signal. Upon another line it indicates that an engine or train is following, but that it possesses no rights, and will keep out of the way. Upon another road it indicates, when carried upon an engine, that such engine or train is wild

White signifies safety, and is a signal to go on.[1]

Green and white is a signal to be used to stop trains at signal stations.[2]

Blue is a signal to be used by car inspectors.

A lantern swung across the track, a flag, hat or other object of any kind, waved violently on the track, signifies danger, and is a signal to stop.

An exploding cap or torpedo clamped to the top of the rail, is an extra danger signal, to be used in addition to the regular signals at night, in foggy weather, and in cases of accident or emergency, when other signals cannot be distinctly seen or relied upon.[3]

or irregular. Upon another line it indicates that the telegraph line is out of order. Upon another line it indicates, when carried upon the rear car, that the train is a regular train. Upon another line it is used at telegraph stations to stop trains for orders. Upon another line when displayed at a switch it indicates that such switch is set for the main track.

1. "At some large stations, where there are lamps showing white lights for other purposes than signaling, which come in the line of the signals, a green light is substituted for a white light on the signal post; but in all such cases trains are to approach and pass through such stations with caution."—*G. W. Ry., England.*

2. When a train does not stop at a station, unless signaled, such station becomes a signal station, so far as that particular train is concerned, but generally speaking, we understand a signal station to mean a small and unimportant place where trains do not stop unless signaled.

3 "Every guard, signalman, engine-driver, gateman, foreman of work, and ganger of platelayers, will be provided with packets of detonators, which they are always to have ready for use while on duty, and every person in charge of a station must keep a supply of these signals in a suitable place, known

The explosion of one of these signals is a warning to stop the train immediately. If the first explosion is followed immediately by a second, the speed of the train need only be slackened, but a sharp look-out must be kept for the regular danger signals. Should a third torpedo be exploded at the regulation distance (600 yards) from the first two, the train must be stopped at once.[1]

A fusee must be used as an extra caution signal. It must be lighted and thrown on the track at frequent intervals, by the flagman of passenger trains at night, or in foggy weather,

by, and easy of access at all times to, every person connected with the station. All the persons above named will be held responsible for keeping up the proper supply of detonators. These signals must be placed on the rail (label upwards) by bending the clasp round the upper flange of the rail to prevent their falling off. When an engine passes over a detonator it explodes with a loud report, and the engine-driver must instantly shut off steam, and bring his engine to a stand, and then proceed cautiously to the place of obstruction, or until he receives an "all right" signal. Detonators must be carefully handled, as they are liable to explode if roughly treated. It is necessary to keep them well protected from damp. At intervals of not more than two months, one from each person's stock must be tested, to insure that they are in good condition."—*Eng. Standard.*

1. Exposure to rain or wet for thirty minutes destroys or impairs the explosive qualities of torpedoes, and, in such cases, too much reliance should not be placed upon them.

"When in snowy weather there is any probability of the detonators being swept from the rails by the brooms attached to the guard-irons of the engines, these signals must not be depended on alone. The guard must not rejoin his train, even though it may be able to proceed, unless some qualified servant of the company can be found."—*Gt. Nor. Ry. Eng.*

The regulations of the Great Northern Ry. of Eng. referred to in this book, were issued in 1856.

whenever the train is not making schedule speed.

A train finding a fusee burning upon the track, must stop, and not proceed until it is burned out.

A semaphore arm extended in horizontal position by day, or a red light by night, signifies danger,[1] and trains must come to a full stop, and not proceed until the signal has been changed by the man in charge, so as to indicate that all is right. When the line is clear and free for the passage of trains, the arm will not be seen by day,[2] and by night a white light will indicate that all is right for trains to proceed. During storms, or in foggy weather, great caution must be observed. If semaphore arm or signal lights can not be plainly seen, trains must be brought to a full stop, and not be allowed to proceed until all is known to be right.

Red signals must be used by telegraph operators and others where the order to stop a train is imperative.

### TRAIN SIGNALS.

Each train, or engine without a train, while running after sunset, or during the day in

---

1. "The danger signal is shown by the arm on the left hand side of the semaphore post standing out from the post."—*Great Western Railway, England.*

2. "The 'all right' signal is shown by the arm hanging down to the side of the post."—*Great Western Railway, Eng.*

foggy weather, must display the white head-light in front of the engine.[1]

Head-lights upon engines must be kept in good order, and always lighted when running after dark, but they must be covered when waiting on turnouts clear of the main track.

Each passenger train, and each through freight train, while running, must have a bell-cord attached to the signal bell of the engine, passing through or over the entire length of the train, and secured to its rear end.

Each train while running must display two red flags at the rear by day. Passenger trains running at night, or in foggy weather, must have two large red lights on the rear platform. Freight and working trains running at night, or in foggy weather, must have three red lights at rear of the train, one being placed on each side of the rear car, near the top, and the other on the rear platform of rear car, or in the cupola, if the car is built with one.[2]

Engines, if alone, running at night or in foggy weather, must carry one red light on rear of tender.

[1]. "The engines carry a white light in front of the passenger trains, and a green light in front of the goods, cattle, mineral, and ballast trains, but north of Doncaster they carry two white or two green lights, to distinguish between goods and passenger trains."—*Great Western Railway, England.*

[2]. "Every train traveling on the line must have a signal lamp attached to the last vehicle, by day as well as by night, except when assisted by an engine in the rear, when such engine must carry the signal."—*Foreign Road.*

A red lantern must be kept lighted and ready for use at night or in foggy weather in the rear car of trains, also upon engines.

All side lights must be covered and the cylinder cocks of engines must be closed when trains are waiting on turnouts, clear of the main track.

Should an engineman observe a train or engine at a stand, on the opposite line to that on which he is traveling, obscured by steam or smoke, he must sound his whistle and approach it very cautiously, so as to be able to stop if necessary.

Two green flags by day, and two green lights at night, carried in front of an engine, denote that the engine or train is followed by another engine or train, running on the same schedule time.[1]

The engine or train thus signaled is entitled to the same schedule rights and privileges as the engine or train carrying the signals.[2]

Two white flags by day, and two white lights at night, when carried in front of an engine, indicate that the engine or train is wild, but

[1] A wild train or a train operated under telegraphic orders, is not allowed under any circumstances to carry signals for a following train.

[2] "A special train to follow is indicated by the preceding train carrying on the last vehicle a red board or a red flag by day, and an additional red tail lamp by night, but as special trains or engines have frequently to be run without previous notice of any kind, it is necessary for the staff along the line to be at all times prepared for such extra trains or engines."—*Eng. Standard.*

the order for wild trains to carry such signals is not imperative.

A yellow flag or lantern carried in front of an engine denotes that the telegraph line is out of order, and the track men of the various sections of road over which this signal is carried must at once examine the telegraph lines, for the whole length of their several sections, carefully and promptly repairing any defects they may discover.

A blue flag by day, and a blue light at night, placed in the drawhead, or on the platform or step of a car, or upon the track, at the end of a train or car, denotes that car-repairmen are at work underneath the said car or cars. The car or train thus protected must not be disturbed until the blue signal is removed by the car-repairmen.

### ENGINEMEN'S SIGNALS.

One short blast of the whistle is a signal to apply the brakes—stop!

A blast of the whistle, of five seconds' duration, is a signal for approaching stations, crossings and drawbridges.

Two long blasts of the whistle is a signal to loosen the brakes.

Two short blasts of the whistle when running, is an answer to the signal of conductor to stop at the next station.

Three short blasts of the whistle when standing, is a signal that the engine or train will back.

Three short blasts of the whistle when running, is a signal to be given by trains, when carrying signals for a following train, to call the attention of trains they meet or pass, to the signals.

Four long blasts of the whistle is a signal to the signalman to return to the train.

Four short blasts of the whistle is the engineman's call for signals at signal boxes, switches, drawbridges and elsewhere.

Five short blasts of the whistle is a call for signals to be sent out to protect the train.

Six distinct blasts of the whistle is a signal to switchman to open the switch so that the engine or train may occupy the side track.

A succession of short blasts of the whistle is an alarm for live stock, or for persons walking or standing upon the track; it is a signal to trainmen of danger ahead.

CONDUCTORS' SIGNALS BY BELL CORD.

One stroke of the signal bell when the engine is standing, is a notice to start.

One stroke of the signal bell when the engine is running, is a notice to stop at once.[1] If,

[1] "Every guard, when traveling, must keep a good look-out, and should he see any reason to apprehend danger, he must use his best endeavors to give notice thereof to the engine-driver. Should a guard wish to attract the attention of the engine-driver, he must, in addition to using the communication, where such exists, apply his brake sharply and release it

after the stroke has been given, and before the train stops, it is found to be unnecessary to stop the train, two strokes will be a signal to the engineman that he may go on.

Two strokes of the signal bell when the engine is standing, is a notice to call in the signalman.

Three strokes of the signal bell when the engine is standing, is a notice to back the train.

Three strokes of the signal bell when the engine is running, is a notice to stop at the next station.

### SIGNALS BY HAND.

The hand moved above the head is a signal to go ahead.

If waved across the body below the head, it is a danger signal or a signal to stop.[1]

suddenly. This operation repeated several times is almost certain, from the check it occasions, to attract the notice of the engine-driver, to whom the necessary "caution" or "danger" signal, as the case may require, must be exhibited."— *Eng. Standard.*

1. "The danger signal 'to stop' is shown by a red flag, or, in the absence of the flag, by both arms held up, thus ☞"

"'Caution,' 'to slacken' is shown * * * by one arm being held up."

"'All Right' is shown * * * by holding the right arm in a horizontal position pointing across the line of rails."—*Gt. Nor. Ry. Eng.*

The two arms extended widely and horizontally, is a signal to back the train.

If both arms are thrown up above the head (touching the hands together), then thrown down by the side, it is a signal that the train is broken apart.

A light swung through a vertical arc (over the head) is a signal to go ahead.

When swung horizontally across the track it is a signal to stop.

When raised and lowered vertically it is a signal to back the train.

When whirled round and round, vertically across the train, it is a signal that the train is broken apart.

**REGULATIONS GOVERNING THE USE OF SIGNALS.**

When upon duty each trainman must carry three torpedoes in his pocket. Passenger trains must also be provided with fusees for use as directed.

Unnecessary sounding the whistle is prohibited, as its excessive use impairs its value as a signal of danger.

The whistle must not be used as a signal for the stopping of a train, except in case of danger, if it can be avoided. It must never be used as the signal for starting a passenger train.

When shifting or moving in yards and at stations, the engine bell should be rung, but the whistle must only be used in cases of absolute necessity.

The whistle must not be sounded while passing a passenger train, except in cases of emergency or danger.

The engine bell must always be rung before starting an engine or train.

When passing or meeting trains on main track or sidings, and when passing through tunnels, or through the streets of cities, towns and villages, the engine bell must be rung.

The engine bell must be rung from a point one-eighth of a mile from every road-crossing, until the road-crossing is passed, and the whistle must be sounded at all road-crossings at grade, where whistling posts are placed.

One stroke of the signal bell while the train is running will be regarded as a warning that the train may have parted, and enginemen will immediately look back and ascertain if such is the case.

When two or more engines are coupled in a train carrying signals for a following train, each engine must carry signals.

When one flag or light (signal) is carried in front of an engine, it must be regarded the same as if two were displayed, but enginemen will be held responsible for the proper display of all the signals required by the rules.

The combined green and white signal is to

be used only to stop trains at the signal stations designated on the schedule. When it is necessary to stop a train at a point that is not a signal station for that train, a red signal must be used.

Switch signals will be arranged so as to show white when the switch is set for the main track, and red when set for the siding, crossing, or junction.

All trainmen, stationmen, switchmen, watchmen, signalmen, operators, track foremen and others whose duties at any time require them to use signals must provide themselves with such signals, and keep them on hand, in good order, ready for immediate use.

## CHAPTER VI.

### CLASSES AND GRADES OF TRAINS.

There are three classes of trains—regular, extra, and wild.

Regular trains are those that are specifically enumerated on the time table.

Extra trains are those that are following regular trains under signals; they possess all the rights of regular trains.

Wild trains embrace all other classes, including those running under special orders or otherwise. They are sometimes called irregular or special trains.

While the grade of trains will vary upon different roads,[1] their importance may be stated generally in the following order:

Passenger:— Express, mail, accommodation, and way.

Freight:— Stock, through, and way.

Construction and wood trains.

All trains will be graded on the schedule in the order of their preference. A train of an inferior grade must, in all cases, keep out of the way of a train of a superior grade.

---

[1]. There are usually only two grades, viz.: Passenger and freight.

## RIGHTS OF TRAINS.

Trains going west have the right of track over trains of the same or inferior grade going east for thirty minutes beyond their schedule time, after which they lose their rights against eastward bound trains of the same or superior grade, and must thereafter keep out of the way. Trains going west will at a meeting station wait five minutes for the expected train, and will then proceed, keeping five minutes behind schedule time until the train is met, except that trains of a superior grade will not wait for trains of an inferior grade. The five minutes is allowed for possible variation in watches, and must not be used by either train.[1]

[1]. This rule is, of course, intended for a single track road, and is based on the supposition that the trains going west have the right of road. It may be made to read in any other direction desired. According to this rule, if the train going west was delayed thirty minutes, the train going east would wait that length of time at the meeting point, after which it would proceed on its way, keeping, however, thirty minutes behind its schedule time, until the delayed train was met. The time which trains must wait varies upon different roads, and sometimes upon different divisions of the same road; thus upon one division they will wait thirty minutes, while upon a neighboring division they will be required to wait an hour.

In the event a company owning a double track road, should, for any reason, be compelled to restrict itself to the use of one track, trains in one direction should have the right of track over trains in an opposite direction, the same as provided for single track roads. In other words, all the peculiarities of operating trains upon a single track road would be enforced. In England, special provision is made for operating a single track in case of a break, the trains are conducted over the line under the immediate direction of a pilot guard, and

Should a train having the right to the road be directed not to leave a station until a specified time, unless another train has arrived, the train so held will wait the usual five minutes for possible variation of watches before proceeding, if the train does not arrive by the time specified.

When a train has orders to run regardless of no train is allowed to pass over the track unless the pilot is personally present on such train; or if there are two or more trains following, he accompanies the last, the forward trains carrying his order to proceed. This order they deliver to the agent at the end of the single line.

Upon many roads trains are ordered to leave the starting point on time, whether trains of the same or inferior grade that are due or past due have arrived or not. In such cases delayed trains are instructed to keep out of the way without reference to the 30-minute rule.

"Trains of a class will start on their time from each end of the road, although a train may be due from the opposite direction. All westward bound trains (trains from Blank) have the right to the road against all eastward bound trains, for one hour after their own time, at any station, per table. After that hour the right to the road belongs to the eastward trains; but no eastward train must leave any station (until the westward train, which was the cause of the delay, has been passed) for not less than one hour after its own time, per table. After passing the delayed train, it can make up what time it can safely. It must be clearly understood that this eastward train which, after an hour's delay, is entitled to the road, has not acquired this right against any other train than the one which was the cause of the delay. This rule is not intended to give any rights to a train of an inferior class against a train of a superior class; but it is only to affect the trains of the same class in regard to each other."

"Westward bound trains of the same class are entitled to the main track at the turnouts, but will take the side-track when arriving in time to do so, if it is known that a train has to be passed at such station, except at side-tracks having but one opening, when the train will enter which can do so head first."—*Regulations Illinois Road*, 1853.

a specified train, it gives the train under such orders no rights over any other train.

Special orders for moving trains are for the persons to whom they are directed, and other persons must not use such orders as authority for moving their trains.

Upon a single track road, when a train is twelve hours or more behind its time, as per schedule, it thereby loses all its rights to the road against all kinds of trains, and can afterwards only proceed as an extra or wild train by special orders.[1]

In case of accident to the engine of a train of superior grade, the conductor of such train may take the engine from the train of an inferior grade, and proceed to destination, reporting the fact from the next telegraph station.

---

[1] Until, therefore, a regular train is twelve hours or more late it is only necessary for it, as it proceeds, to keep off the time of regular trains, of the same or superior grade; until the expiration of the time stated, wild trains must keep out of its way. Upon many lines a train does not lose its rights under the regulations of the schedule until it is 24 hours or more behind time.

## HOW TO PROTECT TRAINS WHEN STANDING UPON THE MAIN TRACK, OR WHEN THE TRACK IS OBSTRUCTED. RULE L[1]

*a.* When an accident occurs to a train, and the road is thereby obstructed, danger signals must be sent in both directions from the obstruction to stop any trains or engines which may be approaching.[2] At a point six hundred yards (paces) from the train, one torpedo must be placed on the rail. At a point twelve hundred yards (paces) from the train, two torpedoes must be placed on the rail, three yards (paces) apart. The signalmen will then return to a point nine hundred yards (paces) from the train, and must remain there until recalled by the whistle of the engine, but if a passenger train is due, the signalmen in the direction of such passenger train must remain until it arrives. When recalled, signalmen will remove

[1] Frequent reference is made to rule "L" as we proceed in connection with the duty conductors and others are under of protecting trains against the possibility of accident whenever, from any cause, trains are compelled to occupy the main track beyond the time allotted them, or when, from any other cause, the track is obstructed.

Many of the rules and regulations necessary to the protection of trains on an ordinary double track road are embodied farther on under the head of "Directions applicable only to double track lines."

[2] Upon a single track road, in the event there is no train due coming from the opposite direction, it seems unnecessary that the signals should be sent in advance of a regular train unless it is over twelve hours late.

the torpedoes nearest the train, but the torpedoes located three yards apart must be left on the rail as a signal of caution to approaching trains. As the delayed train moves on, the torpedoes in advance of such moving train should be removed from the rail. Upon double track roads it will not be necessary to send the signals in advance unless the opposite track is also obstructed. When it is necessary to send the signals in advance, the fireman must perform such duty, and if, from any cause, he is unable to go forward promptly, the front brakeman must be sent in his place.[1] When it is necessary for the rear brakeman to go back to protect a train, the next brakeman must immediately take his place on the train and remain there until relieved by the rear brakeman. On passenger trains, the baggageman shall take the place of the forward brakeman whenever necessary.[2]

[1]. Upon passenger trains this duty can very well be performed by the forward brakeman, there being still one man left upon the train to act as brakeman, viz: the baggageman, but upon freight trains the absence of two brakemen would perhaps leave the train without adequate force.

[2]. "In case of any detention, a man must be sent at least one hundred rods backwards and forwards, to warn any approaching train, until the danger is over. In the night this must be done by swinging a lantern across the track."— 1853.

"In case of a collision, it will be assumed, as a rule, until very clearly proved to the contrary, that the conductors and enginemen of both trains have neglected some of the many precautions, whether written or not, which are necessary to the safety of the road."—*Regulations N. Y. Road*, 1863.

*b.* "Should the distance of twelve hundred yards fall within a tunnel, or close to the mouth of a tunnel nearest to the obstruction, or in any other position where, owing to the formation of the line, or some other circumstance, the engine-driver of an approaching train or engine would be unable to obtain a distinct and distant view of the signal, then the signal must be exhibited at the end of the tunnel farthest from the obstruction, or at such a distance over and above the prescribed distance of twelve hundred yards as may be necessary to insure the engine-driver obtaining a good and distant view of such signal."[1]

*c.* Where a mixed gauge is used, torpedoes must be placed on each rail, both for broad and narrow gauge trains.

*d.* When, from any cause, a train is unable to proceed at a greater speed than four miles an hour, the signalman must go back twelve hundred yards, and must follow the train at that distance, using the proper danger signals, so as to stop any following train, until assistance arrives or the train is switched.

*e.* When a train is stopped upon the main track in consequence of the signals referred to in this rule, the conductor thereof must in turn protect his train with signals, in the manner described, from any train that may be following him,

1. English Standard.

thus relieving the signalman previously upon duty.[1]

*f.* Should anything occur to detain an engine, not attached to a train, upon the main track, it must also be protected by signals in the manner described.[2]

*g.* In the event of any obstruction or accident to the line, not expressly provided for in the foregoing, from the destruction of bridges or culverts, broken rails, washing away of the track, or from any other cause, signals must be placed in both directions, so as to warn approaching trains. These signals must be placed in the manner and form described.

*h.* In the event of any obstruction or accident to the track, as contemplated by this rule, notice of the same must at once be sent to the Superintendent from the next telegraph station; also to the nearest agents or flagmen in each direc-

---

[1] "He (the signalman that is relieved) must tell the guard of such train as he stops what has happened, and ride on the engine, so as to point out to the driver where he left his own train, and tell him the particulars under which he had been obliged to stop the following train."—*Great Northern Railway, England.*

[2] While the instructions contained herein provide specifically for *trains*, they are also, in many cases, intended to cover engines running without trains; in many instances the rules are so worded as to cover both trains and engines; but whether both are mentioned or not, those cases where both are intended will be obvious to the reader. When it is desired to apply a rule to engines that refers, herein, only to trains, but properly applies to both trains and engines, the word conductor, wherever used, should give place to engineman (unless there is a conductor in charge), and engine should be substituted for train.

tion from the accident; but the first duty of employés is to protect approaching trains from any possibility of disaster in consequence of the obstruction.[1]

*i.* In the event any accident occasion the obstruction of, or be dangerously near to, any track used by trains moving in the opposite direction, signals must be placed upon such line, and it must otherwise be protected in the manner contemplated by this rule.

*j.* When a passenger train is delayed at any of its regular stopping places more than five minutes, it must be protected with signals in the manner described.

*k.* Should a train or engine stop at any unusual point on the road (*i. e.* at any point that is

---

[1] "When an accident or obstruction of any kind occurs on any part of the line, it must be immediately reported by telegraph, or by the most expeditious means, to the next station or signal box on each side of the place where the accident has occurred, so that notice may be given to the engine-drivers and guards of approaching trains; also to the heads of departments, to the locomotive station where the breakdown vans for the district are kept; to the district superintendent and the traffic inspector for the district, and to the inspector of permanent way. It must also be reported by telegraph to those stations where the starting of other trains is liable to be affected by the delay caused by the obstruction."

"In conveying intelligence of, or in summoning assistance to, any accident or failure, a platelayer (section-man) must be sent as quickly as possible to the next gang in each direction, from which a platelayer must in like manner be sent to the next more distant gang, until information of the accident has by this means reached the nearest station in each direction, and the necessary assistance has been obtained, the platelayers of each gang proceeding without loss of time to the place at which their services are required."—*English Standard.*

not a regular stopping place for such train or engine), it must be protected as directed.

*l.* When a freight train stops at its regular stopping places where it can be plainly seen at a distance of at least one-half mile, danger signals must be placed not less than one hundred yards in each direction, and as much farther as may be necessary to insure stopping any train that may be approaching, but if the train can not be plainly seen at a distance of at least one-half mile, signals must be sent out not less than six hundred yards, always bearing in mind, that if from any cause the train should be detained, so as to come within twenty minutes of the time of a passenger train, it must be governed strictly as provided by the requirements of this rule, as already recited.[1]

[1] This section, in force upon one of our great railroads, seems unnecessary except upon a double track road where freight trains move without much, if any, reference to the rights of other trains under the schedule.

It is impossible that signals should in all cases be sent out as directed at the various regular stopping places of freight trains. To do so would require an enormous train or station force, and besides, if the freight train is not trespassing upon the rights of other trains, such precautions are unnecessary. If it is in the way of trains having the right to the track, then the precaution provided by this rule is necessary, otherwise it is not. The regulations of trains require that officials in charge of extra or wild trains, or delayed trains of inferior grade, must approach stations carefully, expecting to find other trains at such stations. If trains of an inferior grade trespass upon the rights of trains of a superior grade, then they should be protected in the manner provided. Upon a double track road it would not of course be necessary to send the signals in advance, as provided in the rule, unless the opposite track was obstructed.

We find the following rule, in the regulations of a prominent road, worthy of incorporation here:

*m.* When it is necessary to cross over to the opposite track upon a double track road, or to protect the front of the train from any cause, a signal must be sent ahead as directed.

*n.* If freight trains are, at any time, obliged to keep the main track in passing passenger trains, signals must be sent twelve hundred yards, in the direction of the expected train, to give suitable warning for it to approach carefully; the conductor of the freight train must see that the switches are right for the passage of the approaching train.

*o.* Those in charge of switching engines are required to exercise great care to prevent accident occurring from the obstruction of the main track.[1] Engines or cars must not be permitted to stand upon the main track, except when switching within the limits of the various

---

"Should it be necessary for a first-class train to occupy the main track at a station or turnout, in the time of any train of the same class, which by the time-table should either stop or pass any first-class train at such station or turnout, no signal shall be given to such approaching train, but it must be distinctly understood that when any train occupies the main track at any station or turnout, in the time of any other train of the same class, which by the time-table does not stop at such station or turnout, the proper signal must be sent out to prevent accidents."

[1] "When any train or engine is shunting from one line to another after sunset and in foggy weather, the head and side lights of the engine must be reversed so as to show red against any other train or engine traveling on the line of rails obstructed by the train or engine so shunting. Shunting engines employed exclusively in station yards and sidings must, after sunset and in foggy weather, carry both head and tail lamps showing a red light."—*Eng. Standard.*

yards. When it is necessary to use the main track at any other point, signals must be placed for the protection of approaching trains as required by this rule.[1]

*p.* Should any vehicle in a train be on fire, the train must be stopped, and the conductor must protect it in the manner required. The brakeman or fireman must detach the cars in the rear of those on fire, and the burning cars must be drawn forward to a distance of fifty yards at least, and then be uncoupled, and left until the fire can be extinguished, to effect which every effort must be made.

*q.* Immediately upon the discovery of a signal of danger, enginemen must sound the whistle for brakes as an evidence that the signal has been observed.

*r.* In the event of accident to trains, the persons in charge thereof have the right to call upon sectionmen and others for such assistance as they may require.[2]

1. "No train may shunt on the main line unless absolutely necessary; and a train must be detained at a station where there is a long siding, so as to allow the following train to pass, rather than send it on with a chance of having to shunt on the main line."—*Gt. Nor. Ry. Eng.*

"Guards performing shunting operations at sidings must, in all cases, take care that the vehicles are left clear of the main line, and within the safety points and scotchblocks, and that the points fall properly, and the scotchblocks are replaced across the rails after the operation is completed."—*Eng. Standard.*

2. "In cases of accidents or emergencies requiring such exercise of authority, the conductor or engineer is empowered

*s.* When it is necessary, while switching, or at any other time, to leave a car or portion of a train on a grade upon the main track or elsewhere, the brakes must be set and the wheels securely blocked.[1]

*t.* When it is necessary to back a train (*i. e.* when it is necessary to move it in a contrary direction upon the line) danger signals must be sent not less than one mile in advance of the moving train. A train must only be backed to the first siding; while it is in motion the whistle must be sounded at short intervals. The speed of the train must not exceed four miles per hour, so that the signalman may be able to keep the required distance in advance.

*u.* When a train is run backward, the conductor must station himself on the rear car, in a position so conspicuous as to perceive the first

to summon any person or persons in the employ of the company, by night or day, to render assistance to a disabled train or engine, and any person neglecting or refusing to obey such summons will be discharged."—*Regulations N. Y. Road*, 1854.

1. " When, from any cause, a goods train has been brought to a stand on the main line, where the line is not level, and it is necessary for the engine to be detached from the train for the purpose of attaching or detaching wagons, the guard must, before the engine is uncoupled, satisfy himself that the van brakes have been put on securely, and, as an additional precaution, must pin down a sufficient number of wagon brakes, and place one or more sprags in the wheels of the wagons next to the rear brake in the case of an ascending gradient, and of the foremost wagons in the case of a descending gradient, so as to prevent the possibility of the wagons moving away. The number of sprags must be regulated by the steepness of the gradient, the number of wagons, their loads, and the state of the weather and rails."—*Eng. Standard.*

sign of danger, so that he may give immediate signal thereof to the engineman. The trainmen should be placed so as to facilitate this.[1]

### WHEN TRAINS BREAK IN TWO.

When a train breaks in two, the person who discovers it must signal to the other men on the train, as directed in the code of signals, repeating the signal several times, or until sure they have been observed.

The forward part of the train that is broken in two must not stop until the engineman is sure that the rear part of the train has stopped.

When entirely certain that the rear part has stopped, the forward part may stop; and, after sending back a signal, it will move slowly back to the rear part of the train; but not until a signal to back up has been received from the conductor of the train, who must be very careful not to give such signal unless the rear part is standing still.

If the engineman of the train can not make sure that the rear portion of the train has stopped, he will proceed to the first siding where he will leave his train, and after waiting twenty minutes, he will signal his engine back to the rear portion of his train, presuming that

---

[1] "Whenever it becomes necessary to back a train to a station, it must be done with great care; and, upon obscure parts of the road, a man must be kept constantly in advance of the rear end of the train." 1863.

it is still in motion, and taking great care not to collide with it.

As soon as the men upon the rear portion of the train discover that it has broken apart, they will stop it, and protect the rear by the usual danger signals, as provided by rule "L."

If a following train reaches the detached part before its engine has returned from the siding, the following train will push the detached portion very slowly toward the siding, sending forward a signal twelve hundred yards in advance, and proceeding with great care, expecting to meet the returning engine.

If any train breaks into more than two parts, the rear part must be stopped first, then the part next forward of it, and so on, using great care not to stop any part so as to permit a following portion to collide with it. When stopped, each portion must, if possible, be protected by signals, but the rear of the last section must be protected in any event.[1]

[1]. "Should any part of the train become detached when in motion, care must be taken not to stop the front part of the train before the rear portion has either been stopped or is running slowly, and the rear guard must promptly apply his brake to prevent a collision with the front portion. There may be cases requiring the train to be stopped, owing to the failure of, or accident to, some part of it, when the prompt exercise of judgment and skill is necessary to decide whether to stop quickly, or otherwise. If the engine be defective, the sooner the train can be stopped the better. If any of the vehicles be off the rails, the brakes in the rear must be instantly applied, in order that by keeping the couplings tight, the disabled vehicle may be kept up and out of the way of the vehicles behind,

## TRAINS RUNNING WITH CARE.

Conductors and enginemen are held equally responsible for the violation of any of the rules governing the safety or speed of trains. They are expected to take every precaution necessary to the protection of their trains, whether provided for by the rules or not.

Trainmen must take into consideration the state of the weather, the condition of the track, and the weight of the train.

Trains will run with great care during and after severe rains, and must reduce their speed when the track is in bad order, or when crossing long bridges or trestle-works.

Trains of every description must approach with care places or yards where engines use the main track in switching.

Stations and switches must also be approached with care.

Upon a single track road when an order is given a train to proceed with caution, keeping a careful look-out for a particular train, it is the duty of the conductor in such cases to send

until the force of the latter is exhausted, it being desirable in such cases that the front portion of the train should be brought slowly to a stand. The application of the front brakes might, in such cases, result in further damage, and they should only be applied when the disabled vehicles are in the rear of the train. In all cases the application of brakes behind a disabled vehicle will be attended with advantage."—*Eng. Standard,* signals in advance as the train approaches curves and obscure places in the track.

In all cases of doubt or uncertainty, trainmen and others should take the safe course and run no risks.

### TRAINS MUST STOP.

Whenever one passenger train is to meet another passenger train at a station, whether at a regular meeting point or at a point designated by a special order, both trains must come to a full stop between the switches at the place of meeting.

Engines with or without trains must come to a full stop within four hundred feet of railroad crossings at grade.

Unless otherwise ordered, trains must be brought to a full stop before crossing drawbridges, and must not thereafter proceed until the signal to go ahead is exhibited.

Trains must approach the end of double track and junction switches at reduced speed, and come to a full stop unless the switches are plainly seen to be right.

### TRAINS MEETING OR PASSING EACH OTHER.[1]

Where trains are to meet each other, the train having the right to the road shall occupy

---

[1] "Long sidings are provided at the principal stations on the up and down lines, to enable the goods and coal trains, etc., to be passed by the passenger trains; the sidings must always be kept clear for this purpose; they must not be used as 'lay byes,' for the ordinary work of the stations."—*Gt. Northern Railway, England.*

the main track, excepting when there are special orders to the contrary, or it shall be impracticable thus to pass, in which case sufficient precaution shall be used to prevent accident or unnecessary delay.

The train going on the side track, must take the switch at the nearest end, instead of running by and backing on, except when this is impracticable, in which case the train must be sufficiently protected by signals before running by the station to back on to the siding.[1]

Upon arriving at a place where a particular train is to be met, care must be taken by trainmen to identify such train; in other words, they must not proceed until the right train has arrived.

When a train is not required to stop at a meeting or passing point with another train, it must, at night, or in foggy weather, approach such point with caution, and at reduced speed, being kept under control until satisfied that the opposing train is clear of the main track, and that the switches are properly set.

The conductor of a slow train must report to the Superintendent immediately on arrival at a station, where, by the schedule, he should be overtaken by a faster train of the same class,

[1]. It should be understood that wherever reference is made to the meeting of trains at stations or sidings, such reference implies a single track road, unless otherwise specially mentioned.

in the event the latter does not arrive on time. The conductor of the slow train must not proceed until the faster train passes, without special orders from the Superintendent.

When a freight train is overtaken and passed by one section of a train carrying green signals for other trains, it must wait until all the sections of such train have passed, unless otherwise directed by special order.[1]

Freight trains will be governed by this rule in starting from terminal stations, and in the application of this rule, terminal stations will be considered the same as other stations on the road.

If a way freight train falls behind its time, as fixed in the schedule, it will not yield the road to a following freight train, with which it has no designated passing point, until overtaken by it; but the way freight must be protected by signals from all chance of a rear collision, and will yield the road at the first station after the following train has overtaken it.

### TRAINS APPROACHING STATIONS.

Trains must approach stations and yards where switching engines are located, with extreme caution.

When approaching stations and sidings, en-

---

[1] Or, in other words, it must not proceed until all the *extra* trains have passed.

ginemen must observe whether the switches are set right, and must always be on the lookout for signals.

Enginemen of delayed trains, or trains moved by special order, and of all extra or wild trains, will approach stations with extreme caution upon the supposition that another train will be overtaken or met; or that the main track will be obstructed or occupied.

Enginemen will carefully approach stations at which they ought to meet or pass trains.

Trains approaching stations where a passenger train is receiving or discharging passengers must be stopped before reaching such passenger train, and will not go forward until it moves ahead or signal is given to the first mentioned train to move on.[1]

---

[1]. Permanent danger signals are erected in both directions from stations, by many roads in this country. They are in common use in Europe. These signals are displayed when a train is at a station receiving or discharging passengers, or whenever the track is for any reason obstructed, or the switches are turned. When these signals are displayed, enginemen of approaching trains are required to advance cautiously until otherwise ordered. For the purpose of protecting a train from trains that may be following it, these station signals (or semaphore arms or lights) are not lowered until a specified time after the departure of the train.

The wisdom of protecting trains with permanent or stationary signals, where the business of a line warrants it or its receipts will permit of it, can not be too highly commended.

"Should a train be approaching, stopping at, or leaving a station, on the opposite line, or should shunting operations be going on, he must, on approaching and whilst passing, sound the engine whistle. The whistle must also be sounded on entering a tunnel."—*English Standard.*

*Trains and Stations.*

### TRAINS FOLLOWING OTHER TRAINS.

When two or more passenger trains are running in the same direction, they must keep not less than fifteen minutes apart. And trains that are found violating this rule must be signaled and held until the fifteen minutes has expired. With this exception: a way passenger train making all the stops may follow an express or mail passenger train making no stops, within five minutes, but it must proceed with great caution until the express or mail train is fifteen minutes ahead.[1]

A freight train or engine must not leave a station to follow a passenger train until ten minutes after the departure of the passenger train.[2]

Freight trains following each other must be

---

[1] "Where the block system is not in operation, no train or engine must be allowed to follow any other train or engine on the same line, within five minutes.

"Where the line is not worked under the block system, no passenger train must, during foggy weather or snow storms, follow a goods train, nor must a fast goods train follow a stopping passenger train from a station, nor pass a signal box where trains are ordinarily signaled, within fifteen minutes, nor even then, until the engine-driver has been properly warned of the time of the departure of the preceding train, and where it will next stop."—*English Standard.*

[2] "No detached engine shall be run behind a passenger train, within three miles, and any train following another shall always keep two miles in the rear, and proceed with great caution."—1854.

kept not less than five minutes apart, except in closing up at stations or passing places.[1]

Any train following another train or engine must proceed with caution, keeping at least one mile in rear of it, and must approach all stations and fuel places with care, expecting to find the preceding train taking fuel or water at such station, whether it may be a stopping place, as per schedule, for that train, or not.[2]

When one or more trains are followed, such train, or trains, must never be stopped between stations where the view from the rear of the train is not clear for a distance sufficiently great to stop a train after it has come in sight.

When following other trains, the engineman and others must keep a sharp lookout for the train immediately preceding them, especially when running around curves and approaching stations.

In the event that one or more trains are united, and run as one train, notice of the fact must be given agents, also the conductors and enginemen of trains that are met or passed.

---

[1] "Freight trains will be run in convoys of two or more trains on the same time. Conductors and enginemen will be held responsible to see that the necessary signals are carried." —*Southern Line.*

[2] This is in a certain sense supplementary to the rules directing how many minutes shall elapse between trains of various grades moving in the same direction.

The Superintendent should be advised at the first telegraph station of the consolidation of the trains.

### KEEPING OFF THE TIME OF OTHER TRAINS.

A train of an inferior grade, running ahead of a train of a superior grade, must keep twenty minutes off the time of such superior train.[1]

Except when otherwise specifically provided, wild trains must keep twenty minutes off the time of passenger trains, and ten minutes off the time of freight trains.

A passenger train must not leave a station, expecting to meet, or be passed at the next station by a train having the right of track, unless it has full schedule time to make the meeting or passing point.

A freight train must not leave a station, expecting to meet, or to be passed at the next station, by a train having the right of track, unless it can make the meeting or passing point without exceeding its maximum speed, and occupy the siding, if necessary, before the time required by rule to clear the opposing or following train.

A freight train, which, according to the sched-

---

1. "Trains of an inferior class, moving in the same direction with trains of a superior class, must get out of their way, by going into the nearest siding."—1863.

ule, should be overtaken and passed at a station by another freight train, must keep off the time of the train which should pass it.

It must be understood that a train not having the right to the track must be entirely clear of the main track before the time it is required by rule to clear an opposing train, or a train running in the same direction; if from any cause it should fail to do so, signals must be sent immediately, as provided by rule "L," already given, for the protection of trains standing upon the main track.

When a freight train meets a passenger train on a single track road, the freight train must occupy the siding, and clear the passenger train ten minutes.

### DELAYED TRAINS.

Upon a single track road, in the event a train or engine is delayed between stations and loses its right to the road, the conductor of such train (or in his absence the engineman) must, when the train or engine is ready to move, send danger signals not less than one mile in advance in the direction in which the train or engine is to be moved. The delayed train or engine must only run to the next siding, and while in motion the engineman will frequently sound the whistle, and will not exceed a speed of four miles per hour, to enable the signalman to keep the required distance in advance.[1]

When, from any cause, a train is unable to proceed at a greater speed than four miles an hour, the rear brakeman must go back twelve hundred yards, and must follow the train at that distance, and use the proper danger signals to stop any following train.

In the event a train is delayed by accident or otherwise between stations, and another train having the right to the road approaches (no matter which way it may be going), and the train having the right to the road can not pass the delayed train, then the latter will proceed to the first siding in advance, carrying signals for the following train. At the first siding it will allow the train having the right to the road to go ahead, after which time both the trains will be governed in all respects as in other cases where one train is met or passed by another.

In extreme cases, in which enginemen find it impossible to make their time in running to stations at which they should by schedule meet another train, they may disconnect their engine,

---

[1]. In the event a delayed regular train has time to reach the first telegraph station ahead without trespassing upon the time of another regular train, then, in that case, it has not lost its right (unless it is twelve hours late), and it may proceed directly to such telegraph station without being signaled as directed above.

leaving the train under proper danger signals, as required by rule "L," and run to the next station and notify the approaching train, and then return after their own train.[1] But before proceeding to carry out this rule the engineman must have the written authority of the conductor to detach the engine and proceed as directed.

When a train is delayed it is the duty of agents and switchmen to report the fact to trains that may be following when the latter stop at their stations.

When a train is more than fifteen minutes late, the conductor will report the cause of the detention to the Superintendent at the first telegraph station.

### EXTRA TRAINS.

An extra train, following a regular train and properly signaled by it, must always be taken and considered to be a part of and to have all the rights of the regular train, and no more, and the conductors and enginemen of other trains must so regard it.

An engine of a regular train must not carry a signal for any train, excepting of its own grade, unless in such cases as are herein specifically provided for.

---

[1]. This rule is provided for those extreme cases where, from some sudden and wholly unexpected cause, a train becomes stalled, or is unable to make the meeting point, or back up to the station that it has left.

When it shall become necessary for a train of an inferior grade to follow a train of a superior grade (as an extra), then such following train shall for that time be taken to be of the same grade with the preceding train.

In case a following train is delayed and can not keep up with the signals, it must not consider it has the right to follow the signals against trains having the right of the road, though the train carrying signals for it may have orders to run to a certain point against a train having the right of track; but the following train, when unable to keep up, must keep back and off the time of all trains having right of track, without special and separate orders.[1]

When a train is ordered to carry signals for an extra or following train, the conductor and enginemen of each of the trains affected by the order must be severally notified. It is the duty of conductors of trains carrying signals to notify conductors whom they meet or pass of the fact. They must also notify agents and switchmen at places where they stop.

It is the duty of trainmen and others to care-

---

1. In other words, the order to run to a certain point does not cover the extra or following train unless the latter is specifically mentioned.

"If the following train should fall so far behind as not to be distinctly seen by the forward train at the time of its leaving any station short of the one named in the notice, it must be distinctly understood by all that the flag or lantern will not be carried farther for them."—*Regulations* 1853.

fully observe whether signals are carried by passing engines.

It is the duty of conductors to assure themselves that signals for extra trains are properly placed and secured.

When an engine is carrying signals for another train, the attention of trains that are met or passed (including construction and wood trains) must be called to such signals by three short blasts of the whistle, as provided by the signal code.

When an extra train is following another train, it must be kept near the train ahead on approaching a station where a train is to be met, in order that the opposite train may have as little detention as is consistent with perfect safety, but in all other cases the distance between the two trains must never be less than one mile.

Conductors of trains carrying signals for extra trains must, on arriving at the station beyond which the signals are not to be carried, notify the agent of the fact, and such agent must give notice thereof to such conductors and enginemen as may reach his station subsequent to the arrival of the train carrying the signals, and previous to the arrival of the trains signaled by it.[1]

---

[1] "The guard of the train preceding the special train is required to see that the tailboard flag, or extra lamp, is removed when no longer wanted, and he must inform the per-

When an engine or train leaves a point to which it has carried signals for a following train, before the following train has arrived at such point, the conductor must notify all trains that he meets until he reaches the next telegraph office, when he will report to the Superintendent that he has withdrawn the signals.

A telegraphic or special order directing the movement of a train, includes only the train specifically mentioned in such order, and must not be considered to cover a train that is or may have been keeping it company, unless such train is particularly mentioned.

When two or more trains are running in company, upon the time of one train and the forward train cannot, from disability of engine or other cause, make time, it will run upon a side track, and let the following train go ahead.[1] The conductors and enginemen must, in such cases, see that the train which takes precedence carries the proper signals, and all special orders affecting the movement or safety of either train must be exchanged. Conductors must report the occurrence to the Superintendent at the first telegraph station; they must also notify all trainmen they may meet and the agents at stations as well.

son in charge of each station at which he stops of the description and destination of the train that is following."—*Eng. Standard.*

[1]. The inference is that these trains are of the same grade. It would be impracticable otherwise.

No engine or train shall carry signals for an extra engine or train without orders from the Superintendent, except as provided in the following rule: Should a train be held by another between telegraph stations, the conductor of the train thus detained may require the first regular train passing him, bound in the same direction, to carry signals for him to the next telegraph station, on his arrival at which he must report to the Superintendent for orders; but the conductor of a freight train shall not have the right to have signals carried by a passenger train, in case, at the next telegraph station, or at some intervening place, said passenger train should pass a train of its own class, nor in any case, unless the freight train is in readiness to follow immediately.

A train signaled by another, in accordance with the foregoing rule, would possess exactly the same rights as an extra train, already described.

"When a train is held between telegraph stations and can not proceed, except under the protection of some other train, and there is no train passing (without great delay) by which it may be signaled, except a wild train, the train held may proceed immediately in advance of such wild train to the first telegraph station, at which place it must get out of the way. But those in charge of the delayed train must notify agents and signalmen, also the trainmen they meet, that they are running irregularly in advance of a wild train."[1]

Whenever it shall be necessary to send an extra engine over the road, it must in all cases precede and run on the time of some regular train; it will be entitled to all the rights thereof, and shall carry proper signals therefor. In such cases the regular train shall run five minutes behind its schedule time.[2]

### CONSTRUCTION AND WOOD TRAINS.

When a construction train is going to or coming from work it must proceed with the utmost caution[3]; never risking the safety of trains, and it must never be on the road within ten minutes of the running time of passenger trains. Neither shall it be on the road within ten minutes of the running time of freight trains, except when the points between which it is working are not more than three miles apart.

When at work on a section not extending over three miles from siding to siding, or when special permission is given by the Superintendent, the conductor may keep at work in respect to freight trains only, until the arrival of such

1. Old Rule.

2. When it is desired that the engine running over the road should assist the accompanying train (assuming it to be a freight train) at the various grades, it can be instructed to follow rather than precede. But an engine should never be allowed to follow a passenger train.

3. They must know before starting that all trains that are due have arrived.

trains, but he must in all cases station the proper signals, twelve hundred yards in each direction, when upon a single track, or in the rear only when upon a double track, unless the same is obstructed. The signalman of the construction train must continue on the watch, under all circumstances, until the freight train arrives. On the arrival of the expected train, the construction train must immediately proceed to the siding in advance of such train.

Conductors and enginemen of wood trains will be governed by the same rules as above given for construction trains.

When freight trains are thirty minutes late, construction and wood trains may occupy the main track, but must keep signals not less than twelve hundred yards in the direction of the expected train. Upon the arrival of the expected train, the construction or wood train must at once proceed to the siding.

No construction train will be allowed to run beyond its given limits without orders, except in cases of great emergency, such as accidents to trains, track, or bridges, or when telegraphic communication is broken and orders cannot be received. Under such circumstances, a construction train or engine may run beyond its limits; but such train or engine must not only keep off the time of regular trains, but conductors and enginemen must signal all curves carefully, and look out for wild trains. They will also report the fact of being off their limits, and the reason therefor at the first telegraph station, or if there is no telegraph station, a report must be sent to a telegraph office by the first train, or by special messenger if there is no train.

Two construction trains will not be allowed to run or work within the same limits except in cases of great emergency; in such cases special orders will be given by the Superintendent.

A special order allowing two construction trains to occupy the same limits does not relieve the conductor and engineman of either train from the responsibility of signaling all curves carefully while running, and otherwise protecting their trains properly while at work on the main track, as already directed.

Before leaving stations for the day's work, conductors of wood and construction trains must report to the Superintendent the exact location where they intend to work, and they must not leave the station until they have received a special order or permit from him.

Conductors of construction and wood trains must leave with the station agent at the starting point a memorandum stating where their trains will be operating for the day; this memorandum must be entered in a book to be kept for that and similar purposes. This book shall at all times be open to the convenient inspection of trainmen.

Conductors and enginemen of construction trains are required to stop at all telegraph stations and register time of arrival and departure of their trains, and direction in which moving, and ascertain if any wild engines or trains are on the road; also the limits of any other construction trains that may be at work on the same division of the road.

Conductors of construction trains must keep themselves informed as to the location where wood trains are at work. In the same way the conductors of wood trains must keep themselves advised as to the location of construction trains.

When a limit is given a construction train, it will only embrace the hours from 4:30 A. M. to 8:30 P. M., and the train must not occupy the main track within its limits before or after the hours specified without special orders.[1]

Upon a single track road, signals, as provided by rule "L" for the protection of trains, must always be placed at a distance of not less than twelve hundred yards on either side of the place where construction or wood trains are at work, and a man must remain with such signals. Upon double track roads, signals need only be

---

[1] "Ballast trains must not work on the main line in a fog, except when authorized under special circumstances."—*English Standard.*

placed in the direction from which trains naturally arrive.

In the case of a double track road, if the opposite track is obstructed, then signals must be placed in both directions.

Conductors and enginemen of construction and wood trains will be held responsible for the strict observance of the rules governing the use of signals, and they will be expected to use every additional precaution which particular circumstances may render necessary.

Wood or construction trains must not have signals carried for them by regular trains, nor must they carry signals for other trains, but circumstances may arise compelling them to follow a regular train carrying signals for another train; in such a case the wood or construction train must carry signals for the train that is following.

### WILD TRAINS.

When regular trains are ordered to leave stations ahead of time, they will be considered as wild trains while running ahead of time.

A wild train or engine must not pass over any portion of the road without special orders from the Superintendent, provided this rule does not apply to engines switching within the limits of the various yards.

Conductors of wild trains must report by telegraph to the Superintendent upon arrival at their destination, and must await his reply before leaving the office.

## THE SPEED OF TRAINS.

The maximum speed given on the schedule for each grade of trains must not be exceeded.[1]

Trains must not arrive at a station ahead of time, nor leave a station before the time specified in the schedule, nor shall they run faster between stations than is required to enable them to reach a station in season to start from it on the specified time, without orders from the Superintendent.[2]

When trains are delayed, the lost time must, so far as possible, be made up by shortening

---

[1] "Special trains, whether passenger, fish, horse, cattle, goods, coal, or otherwise, must be run as nearly as practicable at the same rate of speed as corresponding trains, shown in the working time-table, and of which they may form a part, and the speed of special trains must, in no case, exceed that of such corresponding trains, unless under specific instructions from the Superintendent of the line."—*English Standard*.

[2] "Freight trains may arrive at the stations for meeting, and for wood and water, and to take on freight, ten minutes before the time stated in the time-table."—*Regulations*, 1854.

"It is better for a train to have two minutes too little to spend at a station than one more than is necessary, as stops are tedious to passengers, and slow running is better for the road and machinery; and when tardiness is noticed in the wooding and watering, it should be reported to the Superintendent."—1853.

the stops at stations.¹ No risk must be incurred for the purpose of making up lost time.

Mail trains must not be run at such speed as to prevent the mails being exchanged at all places where there are post offices.

A speed of fifteen miles per hour will pass, approximately, seven telegraph poles per minute.²

### DIRECTIONS APPLICABLE ONLY TO DOUBLE TRACK LINES.

All trains in either direction, when running on a double track, will invariably take the right-hand track.³

On a double track road, when a freight train passes over to the opposite track to allow a passenger train running in the same direction to pass it, if, while waiting, a passenger train in the opposite direction arrives, the freight train may cross back, and allow it to pass; provided,

---

1. "When passenger trains are behind time, the engineer is at liberty to make it up, in whole or in part. with the consent of the conductor, when he can do so with safety."—1863.

"Their trains should be so run as to leave at stations only the necessary time for doing the business of the train, that as much time may be used in running and as little in stops as possible. They will. after attending to their passengers, see that what remains to be done to enable them to leave the station is done in the shortest possible time."—1853.

2. U. S. Road.

3. "The engine-driver must start and stop his train carefully, and without a jerk, and pass along the proper line, which, in the case of an ordinary double line, is the left-hand side of the permanent way, in the direction of which the engine is traveling."—*Eng. Standard.*

the other passenger train is not in sight; and also provided, that danger signals have been sent not less than twelve hundred yards in the direction of the expected train, as provided by rule " L " for the protection of trains.

On a double track road, when it is necessary for a freight train to cross over to the opposite track to allow a passenger train running in the same direction to pass it, and a passenger train running in the opposite direction is due, danger signals must be sent back twelve hundred yards, as already described in rule " L," and the freight train will not cross over until one of the passenger trains arrives. Should the following passenger train arrive first, danger signals must be sent forward (as per rule referred to above), not less than twelve hundred yards in the direction of the over-due passenger train upon the opposite track before crossing over. Great caution must be used, and good judgment is required to prevent detention to either passenger train; preference should always be given to the passenger train of superior grade.

If an obstruction or accident make it necessary to move an engine or train in the wrong direction on a double track road, or to cross over to the opposite track to pass around such obstruction, obstructed trains or engines may do so, but the utmost caution must be used. The conductor of the obstructed train (or in his absence, the engineman), before the engine is moved, will send danger signals not less than one mile in advance, in the direction in which the train is to be moved. The train or engine thus moved must only be backed or run to the next crossing, and, while moving, the engineman will frequently sound the whistle, and not exceed a speed of four miles per hour, to enable the signalman to keep the required distance in advance.

Freight trains, in cases described in the foregoing rule, must clear the time of passenger trains twenty minutes.

Upon a double track road a train that is delayed and falls back on the time of another train of the same grade, does not lose its rights, and will not take the time or assume the rights of another train, except as provided for herein, without orders from the Superintendent.

Upon a double track road, no conductor shall assume the rights or take the time of any other train without special orders from the Superintendent, except as provided in the following rule.

A train overtaking another train of the same or superior grade will not run around it, except the train ahead is disabled by accident, but in such case, the train passing the disabled train will assume its rights and report the fact to the

Superintendent from the next telegraph office. The disabled train will assume the rights of the last train passing it, and report to the Superintendent from the next telegraph office. When the rights of one train are assumed by another train, notice of the fact should be given agents and others at places where the train stops.[1]

It must be kept constantly in mind by trainmen, when occupying the left-hand track of a double track road (*i. e.*, when occupying the wrong track), that they are responsible for keeping out of the way of trains that rightfully belong on such track, and they must in all such cases protect trains with adequate signals as described.

Should a train, which has been telegraphed as having entered a tunnel, not emerge therefrom within a reasonable interval of time, the

---

1. A prominent company having a double track road provides as follows where a delayed train impedes other trains: "Extra freight trains running ahead of regular freight trains can take the time of such regular train when the regular is behind its table time, or can do so when necessary to get over portions of single track. Conductor of such extra must leave written notice for conductor and engineer of regular train informing them that he has then and there taken their time, and availed himself of their rights, in which case he is authorized to make the time of the train under whose rights he is running. It must be distinctly understood that subordinate trains or engines are still subordinate, though an extra freight is running on the rights of a train having priority."

It will be noticed that no provision is made for notifying the Superintendent of the transfer of rights at the first telegraph station, from which it may, perhaps, be inferred that the train that assumes the rights of another continues to exercise those rights until it arrives at its destination.

signalman toward whom the train is approaching must prevent any train in the opposite direction entering the tunnel, through which there is a double line of rails, until he has ascertained that the line on which it has to run through the tunnel is clear.

Should an engineman observe anything wrong on the line of rails opposite to that on which his train is running, he must sound the whistle and exhibit a danger signal to any train or engine he may meet, and stop at the first station and report to the person in charge what he has observed. Should he meet an engine or train too closely following any preceding engine or train, he must sound the whistle and exhibit a caution or danger signal, as occasion may require, to the enginemen of such following engine or train.

Upon a double track road, when a portion of a train is left upon the main line, from accident or inability of the engine to take the whole forward, the engineman must not return for it on the same line except by written instructions from the conductor, but must go on the proper line and cross at the nearest point behind the part left (unless there is a crossing in its immediate front), which he must push before him till convenient to go in front again with the engine. If the engineman finds it necessary to return to the rear portion of his train on the same line, he must, before starting with the front portion, send his fireman back to the conductor to obtain the necessary written instructions authorizing him to do so, and if he give such instructions, the conductor must continue to protect his train in the rear and prevent a following train pushing it ahead, except upon inclines worked under special rules.[1]

[1] "In the event of an accident occurring, whereby one of the main lines is obstructed, the traffic in both directions must be carried on by the other line ; but this must not be done until the following rule is rigidly put in force:

"A pilot engine must at once be procured, and in the event of there not being a pilot at hand, the engine of a goods or coal train must be taken temporarily for the purpose, and written orders having been given, at both ends of the single line, by the chief officer on the spot, that no engine or train be allowed to go on to it without the pilot engine is at the end from which the train is about to start, the district agent, clerk in charge of the principal station near which the obstruction has taken place, or other officer, will proceed to pass the traffic on one line, accompanying the pilot engine backwards and forwards, and directing the arrangements at both ends of the single line. If no pilot engine can be procured, one man, whose name must be given to the person in charge of such contiguous stations or crossings, must be appointed, in writing, to act as pilotman, and he must ride on every train or engine in both directions, and no train or engine must move from the said stations or crossings unless this man is riding with it : and this one man must continue riding to and fro between the aforesaid places until relieved, and a successor named in writing, at the two ends of the single line then being worked."—*Gt. Nor. Ry. Eng.*

"In case of accident blocking or breaking one track and requiring a train to pass along the wrong track, the utmost caution must be exercised, and no train or engine must be permitted to proceed on the wrong line without a memorandum in writing from the person in authority at the spot where the accident had happened, and station agents must be satisfied that such orders have been given and received, that all trains have been stopped until the arrival of the one they dispatched on the wrong track."—*N. Y. Road*, 1854.

### THIRD TRACK, OR MIDDLE SIDINGS.[1]

The middle sidings, or third track, must be used by trains (in either direction) whenever it is necessary to turn out to allow trains of a superior class running in the same direction to pass them.

A half-way post will be placed in the center of each middle siding; trains in either direction may run to the half-way post at a speed not exceeding six miles per hour, but must not run beyond it, except under the protection of danger signals.

When trains pass the half-way post, they must run at a speed not exceeding four miles per hour, to enable the signalman to keep not less than six hundred yards in advance of the train.

When two trains meet on a middle siding, the train nearest the switch shall be backed, keeping a flagman not less than six hundred yards in advance; but when there are crossing-switches in the center of a middle siding, they must be used in all cases when the backing of either train from the siding on to the main track can be avoided.

All trains are required to use middle sidings

---

1. No changes whatever have been made by me in the regulations governing the use of the third track. I have accepted them just as I find them in operation on one of our greatest as well as one of our most carefully managed roads.—M. M. K.

with great care; they must invariably run expecting to meet an opposing train, whether opposing trains are due or not.

### COUPLING CARS.

Care must be exercised by persons when coupling cars.

The coupling apparatus of cars or engines is not always uniform in style, size or strength, and is liable to be broken.

It is, therefore, dangerous to expose the hands, arms or persons of those engaged in coupling cars.

Employés are, therefore, directed to examine, so as to know beforehand, the kind and condition of the drawhead, drawbar, link and coupling apparatus, and are prohibited from placing in the trains any car with a defective coupling. Sufficient time is allowed and may be taken by employés to make the examination required.

Coupling by hand is prohibited in all cases where a stick can be used to guide the link or shackle; and each switchman, brakeman or other employé who may be expected to couple cars, is required to provide himself with a stick for that purpose.

Uncoupling cars while in motion should also be avoided.

*Trains and Stations.* 125

MISCELLANEOUS ORDERS RELATIVE TO TRAINS.

Regular trains will be run in accordance with the schedule, except when otherwise ordered by the Superintendent.

No passenger train must be stopped at a station where it is not timed to call, for the purpose of taking up or setting down passengers, without special authority.[1]

The time indicated in the schedule is the arriving time of trains, except when the time of departure is expressly stated.

Large full-faced figures upon the schedule, opposite a station, indicate that other trains are met or passed at that station.

Trains shall be run uniformly and steadily between stations, and delayed as little as possible for fuel and water, and for the transaction of station business.

Passenger trains shall be drawn, not pushed, except in case of accident or other emergency.[2]

When express or freight cars are hauled in a

---

[1] "All passenger trains are to stop at the stations mentioned on the time bills, whether there be passengers to alight from the carriages or not."—*Gt. Northern Ry. Eng.*

[2] "No engine must be allowed to push a train of carriages or wagons on the main line, unless within station limits, but must in all cases draw it, except under special regulations when assisting up inclines, or when required to start a train from a station. In case of an engine being disabled on the road, the succeeding engine may push the train slowly to the next siding, or cross-over road, at which place the pushing engine must take the lead."—*English Standard.*

passenger train, they must be placed next to the engine.

No train shall start without a signal from its conductor, and conductors must not give the signal until they know that the cars, including the air brake hose, are properly coupled.

At points where registers are kept, or where train boards or indicators are located, it is the duty of those in charge to see that the arrival and departure of trains are accurately and promptly noted thereon, the grade of the train being given in each instance.

When the track is clear, a white signal must be displayed from stations where trains pass without stopping.

Pieces of wood or coal must not be thrown from an engine or train when in motion, lest sectionmen or others be injured thereby.

Flying switches must not be made, except at places or by persons authorized by the Superintendent. In the absence of such authority a switch rope must be used.[1]

No person will be permitted to ride on the engine or tender without an order from the Superintendent, except the engineman, fireman, inspector of engines, and road masters in the

---

1. " Double shunting is strictly prohibited, except when done by engines specially used for the purpose of shunting, and attended by experienced shunters. Fly shunting of empty vehicles against loaded passenger trains, and of vehicles containing passengers or live stock is strictly prohibited."—*English Standard*.

discharge of their duties on their respective divisions, and trainmen, in cases of accident, or whenever necessary.

Employés, when on duty in connection with the train service, will be under the authority, and conform to the orders, of the Superintendent of the division upon which they may happen to be at work.

Mail agents, messengers of express companies, sleeping car conductors and porters, news agents, and individuals in charge of private cars, must consider themselves as employés in all matters connected with the movement and government of trains, and must conform to the directions of the conductors of the trains upon which they may be employed.

Conductors and enginemen are required to compare time daily with the standard time of the company.

In order to insure uniform time being kept at all the stations on the line, to which time is not telegraphed, the following regulations should be strictly observed:

Each conductor must, before starting on his journey, satisfy himself that his watch is correct with the standard clock, and must again compare it, and regulate it, if necessary, at the end of his journey, before commencing his return trip.

The conductor in charge of the first passenger train (starting after 8 A. M.), stopping at all stations on the portion of the main line, or branch over which it runs, must, on his arrival at each station at which there is no telegraph office, give the agent or other person in charge the precise time, in order that the station clock may be regulated accordingly; and, in the event of the time given by the conductor differing from that of the station clock, the latter must be altered to agree.

The agents will be held responsible for keeping their clocks properly regulated in accordance with this order, and must at once report to the Superintendent any serious defects that may occur in their working, in order that the necessary steps may be taken for their immediate repair.

Conductors of trains running at night, upon a single track road, are required to report in person to the operator at every night telegraph office at which they stop.

At night the conductors of freight trains will make and sign duplicate statements (memorandum cards) of the time of leaving each station, and give such statements to the telegraph operator, or, in case there is no operator, to the watchman. When the next train going in the same direction arrives, the operator or watchman will hand the copy to the engineman of such train. Enginemen will be on the look-out to receive such notices as they pass stations. At stations where train registers are kept for the information of trainmen, this rule need not be observed.

All accidents, detention of trains, failure in any way of engines, or defects in the road or bridges, must be reported to the Superintendent by telegraph from the next station.[1]

### THE TRACK.[2]

" The laborers must be in squads of such number and force as the roadmaster may direct, and to each squad there must be a foreman, who must work constantly with his squad, and be held responsible for the faithful and efficient execution of the work under his care."[3]

The safety of life and property requires that sectionmen should be especially vigilànt in foggy weather and during and after storms.[4]

1. "Conductors and engineers are required to report promptly any defect they may discover in the track, to the Superintendent of repairs of track."—1853.

2. Generally speaking, only those rules that immediately affect the movement and safety of trains are embraced herein. "THE ROAD-MASTER'S ASSISTANT," by Huntington, revised by Latimer, C. E. A. A. G. W. R. W., and published under the auspices of the *Railroad Gazette*, contains a very clear and exhaustive statement of the duties and responsibilities of trackmen. It is worthy of the perusal of managers and trackmen. —*Regulations*, 1854.

3. " In each gang of platelayers or men repairing the permanent way, there shall be a foreman or ganger"—*Eng. Standard*.

4. " They must see that after all heavy winds, rains, and other storms, and during the same, the men are out on the road

In no case, except in the most absolute necessity, is a rail to be displaced or any other work to be performed, by which an obstruction may be made to the passage of the trains during a fog or snow storm, and the times for effecting repairs which involve the stopping of trains must, as far as practicable, be so selected as to interfere as little as possible with the passage of the traffic.[1]

"In case of accident to trains the nearest section foreman will at once take his whole ready to render such assistance as may be required, and to give proper warning to the trains, and to repair such damages and remove such obstructions as are necessary.

"In foggy weather, when a train can not be seen at three hundred yards, all the foremen and laborers must leave their ordinary work, and the foreman must range them along his portion of the line, over which they must walk up and down, driving such spikes and keys, or doing such other work as needs attention, and be ready to give notice of danger to the signal-men or the trains.

"They must see that such buckets, axes, and other tools are kept at the bridges and other structures, as to protect them from fire and other damage; and that each squad of laborers is supplied with, and keeps ready for use when at work on the road, white and red flags, lanterns, and torpedoes.

"They must see that all rocks, stones, earth, trees, stumps, and other things that are likely and liable to fall on the track or endanger the trains, be thrown down or removed, and such other measures taken as to insure the safety of the road." 1854.

"Trackmen should appreciate the fact, that the safety of the lives of passengers, and of the property transported over the road, is largely dependent upon their watchfulness and discretion, and that any failure to discharge their duties promptly and thoroughly may result in the destruction of both." 1863.

1. "In all cases, before taking out a rail, the platelayer must have at the spot a perfect rail in readiness to replace it."—*Eng. Standard.*

force to the assistance of the train, even if it is not on his own section.

"In case of a wreck, foremen must at once appoint the necessary watchmen to prevent freight or company's property from being stolen. Said watchmen are to remain on duty until the goods are removed.

"On receiving notice of a wreck or accident they (roadmasters) must at once proceed to the place and take full charge and control of all track forces and construction trains; put the track in condition for the safe passage of trains; and remove the wreck with the quickest possible dispatch."[1]

The gravel or ballast unloaded along the line must be promptly spread upon the track, so as not to endanger the safety of trains.[2]

Fuel, ties, or material of any kind must not be piled within six feet of the main track.

"In lifting the permanent way, no lift must be greater than three inches at once, and then it must be effected in a length of at least twenty yards, in such a manner as not to occasion any sudden change of gradient. Both rails must be raised equally and at the same

---

1. Southern Line.

2. "No ballast must be thrown up to a higher level between the rails than three inches, and it must be thrown as much as possible on the outside of each line and between the two lines, and be replaced as soon as possible. The rails must be kept clear of gravel, ballast, or any other material."—*Eng. Standard.*

time, and the ascent must be made in the direction in which the trains run."[1]

When making repairs that obstruct the track, or jeopardize the safety of passing trains, sectionmen must place danger signals upon the track, as required by rule "L," for the protection of trains.[2]

If the track is in bad order, or if, for any other reason, it is desired that trains should run slowly, green signals must be used.[3]

Sectionmen must keep the fences in good order at crossings and at each side of the track; they must see that all breaks are repaired

---

1. English Standard.

2. "When repairing, lifting the line, or performing any operation so as to make it necessary for a train to proceed cautiously, the foreman or ganger must send a man back at least half a mile, and as much farther as the circumstances of the case render necessary, who must exhibit the 'caution' signal so as to be plainly visible to the engine driver of the approaching train.

"Each gang of platelayers or laborers must be supplied by the inspector of permanent way for the district with two sets of day signals, two hand signal lamps, if working after dark, and a proper number of detonators. Each ganger will be held responsible for having his signals constantly in proper order and ready for use."—*Eng. Standard.*

3. "A green flag, or a green light, exhibited by platelayers, indicates that trains and engines must reduce speed to fifteen miles an hour over the portion of line protected by such green signal. The 'caution' signal must always be exhibited at a distance of at least half a mile from the point where it is required that the speed of trains and engines should be reduced and as much further as the circumstances of the case render necessary."—*English Standard.*

without delay;[1] that cattle guards are kept in repair; that all gates that are found open are closed, and that all bars found down are put in proper condition.[2]

When watchmen are employed, they must walk over the track and carefully inspect the same, at intervals between the passage of trains.[3] It is the duty of watchmen (and switchmen and agents as well) to signal trains that disregard the regulations prescribing the

---

[1]. "Surely, it is far better to stop a hand car and repair a fence than to subject a company to damages for killing stock, with the additional expense occasionally, of a wrecked train.

In a word, men, when passing over a road with a hand car, should be prompt to remedy every defect they discover. It should be a rule never to postpone any work of repairs that can be done on the instant."—*The Roadmasters' Assistant*, p. 118.

[2]. "Gangers must close and fasten all gates they find open, and report the circumstances, in order that the persons who are required to keep such gates closed and fastened may be charged with the penalties."

"The gangers must take care to maintain proper scotches on all sidings requiring them."—*English Standard.*

[3]. "Whenever any person has occasion to walk on the railway he must not walk on either line of rails, but on the right hand side of the line, off the ballast, clear of passing engines or trains."—*Great Northern Railway of England.*

"Gangers must order off the railway all persons trespassing within the fences, and must do their best to obtain the trespasser's name and address. If any trespasser persists in remaining, they must take him to the nearest station and give him in charge of the station master or police there; or (if any police constable be nearer than the nearest station) gangers must give the trespasser in charge of such constable, and at once report having done so to the nearest station."—*Great Western Railway of England.*

time and distance that must elapse between trains that are following each other.[1]

Trackmen must observe the condition of the telegraph lines as they pass over their sections, and in the event the line is broken or obstructed, they will make such temporary repairs as may be required, reporting the circumstances of the case to the operator at the next telegraph station.

"Each ganger is required, in the event of storms or floods, to examine carefully the action of the water through the culverts and bridges on his length of line; and should he see any cause to apprehend danger to the works, he must immediately exhibit the proper signals for the trains to proceed cautiously, or to stop, as necessity may require, and inform the inspector thereof; and until the inspector arrives, he must take all the precautionary measures necessary for securing the stability of the line."[2]

---

1. "The foreman and other men of the squads must look at every passing train, and if they see a train running on the same track, within ten minutes of another train, or anything wrong, they must signal to the engineman with a red signal, and they must report to the trackmaster when an engineman does not obey the signals."—1854.

"Where the line is not worked under the block telegraph regulations, if a passenger train approach within ten minutes of a goods, cattle, mineral, or ballast train, or light engine, the men repairing the line must give the engine-driver of such passenger train a signal to go slowly."—*Eng. Standard.*

2. G. W. Ry., England.

They must see that the ditches are kept open, and that the water courses under the bridges and culverts are not allowed to become clogged or obstructed.[1]

In wet weather, and during and after snow storms, they must use every effort to prevent delay or accident to trains.[2]

Track foremen must carefully inspect every portion of the section under their charge at least once each day.[3]

"Each ganger must, when going over his length of line to examine the keys and fastenings of the rails, have with him a keying hammer and spanners or nut keys, and be prepared promptly to supply keys, nuts, packings, fast-

---

[1]. "They will be particular not to allow standing water upon any part of their line, but keep the ditches open and free at all times, and keep flood-wood away from the culverts, bridges, and water-courses."—1853.

[2]. "Their whole time will be devoted to their duties in the service of the company, and generally their services are more urgently required in bad, inclement weather than at any other time."

"In winter, it is as much their duty to keep the track clear from snow and ice, as far as it is possible, as to keep it in repair. At this season every possible effort should be made to keep the road open, and insure the regularity of trains."—1853.

[3]. "Each ganger must walk over his length of line every morning and evening on week days (except where the engineers consider once each day sufficient, and have laid down such instructions in writing) and where passenger trains are run, once on Sundays, and tighten up all keys and other fastenings that may be loose; and he must examine the line, level, and gauge of the road, and the state of the joints, marking, and if necessary, repairing such as are defective."—*G. W. Ry., Eng.*

enings, or other parts of the permanent way that may be required."[1]

"No wagon or other vehicle employed in the permanent way department must be left in any siding without the wheels nearest to the entrance into the main line being properly scotched and secured."[2]

No notice will be given trackmen of the passage of trains, and they must therefore govern themselves accordingly.[3]

Section foremen must report to the Superintendent any neglect upon the part of trainmen to properly regard danger or caution signals.

Old or unused material of every kind upon the line of the road, or at stations or shops, must be carefully collected and preserved.[4]

1. G. W. Ry., England.
2. English Standard.
3. "On no occasion, except in cases of emergency or of accident, and never at night, or in a fog, or when a train is due, must a trolley be run in the wrong direction, and in such cases the trolley must be preceded at a distance of not less than a mile by a man with a red flag and detonators. In tunnels a red light must always be used."—*Great Wes. Ry. Eng.*

"In the case of a single line, the trolley must be so protected in both directions. No trolley must, in any case, be placed on the line, except by the platelayers and with the knowledge of the ganger, who is responsible for seeing it properly protected and used. No trolley must, under any circumstances, be attached to a train, and all trolleys when not in use must be taken off the rails, placed well clear of the line, and the wheels secured with chain and padlock."—*Eng. Standard.*

4. "They will protect the materials or property of the company (whether new or old) upon their line from depredation, loss or injury, and keep it properly and neatly piled up, ready for use or removal."—1853.

"All luggage, goods, or articles found on the line must immediately be taken to the nearest station, and a report made containing the best information that can be obtained respecting the train from which they may have fallen.[1]

" Trackmen working in a tunnel, when trains are approaching in both directions, must, if unable to reach any recess in the walls, lie down either in the space between the two lines of rails, or between the line and the side of the tunnel, until the trains have passed. The width of the space depends on the construction of the tunnel, with which every man must make himself acquainted in order that he may select the place which affords the greatest safety."[2]

Trackmen must desist from work upon a train approaching, and must not cross over to the other lines, but move to the side of the road, clear of all the lines, to secure themselves from the risk of accident by trains running in opposite directions.

In the event of any fire taking place upon or near the line, employés must take immediate measures for putting it out.[3]

1. Eng. Standard.
"Anything which may have been lost from a passing train, such as a casting, nut, screw, or bolt, or any piece of machinery, piece of freight, baggage, or other matter, they will pick up and carry to a regular station, and deliver to the station-agent."—*Old Rule.*

2. Eng. Road.

3. "Careless firemen frequently throw overboard handfuls of dirty waste, which at any time may be ignited by a spark

Bridges and culverts should be carefully inspected after the passage of each train; but where this is impossible they must be examined daily, or oftener if sectionmen have occasion to pass over them. All defects should be promptly remedied, and in the event sparks, burning waste, fuel or fire of any kind is observed, it should be put out.[1]

"Before removing any traveling crane, the person in charge of it must see that the jib is properly lowered and secured, and so fixed that it will pass under the gauge, and, when it has to be removed by train, it must, when practicable, be so placed that the jib will point towards the rear of the train.

"Whenever a crane is in use whereby the jib, or any other portion of it, obstructs or fouls any line of rails in use for traffic purposes, or whenever, by any possibility, during the loading of round timber, long timber, angle iron, or

---

from a passing locomotive. Fire may be carried thence into the dry grass by the roadside, afterwards into the fence, and soon to the hay stacks, buildings, wood piles, etc."—*The Roadmasters' Assistant*, p. 116.

[1] "When a gang of trackmen engaged at work discover a smoke on a line, they should at once attend to it. It should be a rule at all times never to neglect the least indication that a fire has caught on the line. On more than one occasion expensive bridges have been destroyed owing to a neglect to stop the hand car and remove a live coal of fire dropped by a locomotive, or to put out a fire caused by a spark from a smoke stack lodging in a decayed spot of timber. Some of the worst wrecks on record have been taken out of culverts where a stringer has been nearly burned through."—*The Roadmasters' Assistant*, pp. 116–117.

other articles of great length, the main line may be obstructed, it is incumbent on the person in charge of the loading to place danger signals "[1] as required by rule "L."

### MOVEMENT OF TRAINS BY TELEGRAPH.

Superintendents and train dispatchers are the only persons authorized to move trains by special orders, and but one person on the same subdivision of a road should be permitted to move trains by special orders at the same time.

Before an order is given by telegraph for two or more trains to meet at a given station, the red signal to stop the trains must first be displayed at such meeting point, and until this is done no order must be sent to either train.

When a meeting or passing point is to be made by two or more trains, the order must be definite and conclusive; it should first be sent to the conductor having the right to the road.

If it is desired to give a train the right to run against a passenger train, the order must first be sent to the conductor of the latter, and no order must be given the opposing train until the receipt of a satisfactory reply from the conductor of the passenger train. And in the same way, before giving a passenger train the right to the road, over a train possessing such right, the order should first be sent to the train

---

[1]. English Road.

having the right to the road; when a satisfactory reply has been received from the conductor of such train, then the order may be transmitted to the other train.

A train of an inferior grade must not be directed to move ahead of a regular train of a superior grade, unless it shall have full schedule time (according to the regulation for trains of that grade) to reach the point to which it is ordered, in advance of the time at which the train of a higher grade is due at such point. And in the event of a train of an inferior grade running ahead of a regular train of superior grade, as directed in this rule, can not make schedule time, its conductor, as soon as he discovers such to be the case, must leave a signalman to warn the approaching train, ahead of which he has been ordered to run, and must report to the Superintendent for orders at the next telegraph station. The conductor and engineman of the train of superior grade that is following must be notified in writing of the order directing the train of an inferior grade to proceed, but it must be distinctly understood that such conductor and engineman will not be held responsible for any accident that may occur in consequence of the slow train getting in its way, unless such accident shall have been caused by a disregard of signals or the rules and regulations.

All special orders for the movement of trains, whether sent by telegraph or otherwise, must be communicated in writing.

When a train is abandoned, the order of the Superintendent directing its abandonment must be sent by telegraph to all agents, conductors, and enginemen upon the division.

No train must leave a station to run upon the time of an abandoned train, which, by the regulation, would have the right of road, unless the conductor and engineman of such train have in their possession a copy of the order of abandonment, properly signed and certified to by the operator.

Should a train be held at night at a telegraph station, where there is no night operator, the conductor will call the day operator into the office for the purpose of receiving the orders necessary before proceeding.

To enable trains to move with promptness and regularity, such expedition as is consistent with safety is enjoined upon trainmen and telegraph operators in the transmission and response to telegraph orders.

Conductors and enginemen must not ask for running orders until they are ready to leave the station, and when an order has been received it must be executed without delay, unless otherwise directed.

Safety demands that all persons connected

with the movement of trains by telegraph should use the utmost care and watchfulness.

The rules regarding or affecting the movement of trains must be strictly observed.

Orders must be made plain and explicit, and if not fully understood by those to whom they are addressed, an explanation must be asked for by such persons before taking any action.

After the reception of an order it must be strictly obeyed.

In the transmission of orders by telegraph, no abbreviations will be used, figures being written in full.

At stations where telegraphic orders are awaiting an expected train, operators will display a red flag by day or a red light by night. Each station must adhere strictly to the locality fixed upon as the best for the purpose of displaying the signals, and such place, once selected, must not be changed except for good and sufficient reasons.[1]

Signals must be promptly removed by operators when the object for which they were displayed has been accomplished.

When the signal is shown as provided above, the approaching train will, in all cases, be brought to a full stop (and in such cases it is

[1]. "No new signal must be brought into use, nor any alteration made in the position or use of any existing signal, without the authority of the Superintendent of the line."—*Eng. Standard.*

the duty of operators to see that trains are so stopped), and the conductor must go immediately to the telegraph office to receive and answer such orders as may be awaiting the train. Should the signal have been displayed for some other train, the conductor must, before proceeding, receive from the operator a written release, stating for what train the signal was displayed. Agents and operators will, upon receiving telegraphic orders for expected trains, immediately exhibit the proper signals, as required by the foregoing rules. The signal must not be relied upon exclusively to hold trains. Operators must watch closely for the expected train, using all necessary means to stop it. In case the train, or any part of it, has already passed the telegraph office, although still at the station, the operator's understanding of the telegraphic order to hold the train must not be transmitted to the Superintendent until the engineman and conductor have been shown such order and understand that they are held for orders.

Conductors and enginemen must, in all cases, read the order so as to avoid any danger of misunderstanding it.

All orders by telegraph for the movement of trains must be taken in triplicate by operators. Manifold paper should be used for this purpose, so that correct copies may be secured. Orders will be addressed to the conductor and engineman, to whom they will be read aloud by the operator, before their delivery by him. The conductor must, upon receipt of an order, write out in full his understanding of such order. This understanding must be read by the engineman, and must be signed by both the conductor and engineman. It will then be transmitted to the person giving the order, who will, if the same is correctly understood, reply "O.K.," giving the time. This reply will be endorsed on the order by the operator receiving the same, over his signature. After being endorsed, one copy of the order will be given to the conductor, and one copy to the engineman, the remaining copy being placed on file. Each official must receive the order in person from the operator. No order will be acted upon until the reply "O.K." has been received and endorsed upon the order as described above; or, in other words, conductors and enginemen must not leave a station when directed to run by special orders, without each having the same in writing in their possession, properly endorsed.

After the receipt of an order, should the line cease to work before the reply "O.K." is received, as per the preceding rule, the operator will not deliver such order, but will inform both the conductor and engineman of the occurrence. It is their duty to adopt such precautions as will prevent accident. Trains will not proceed in such cases, except under protection of signals, until all doubt has been removed.

When an order is sent to a train which may be carrying a signal for a train, such order will not cover the train that may be following, and in no case will the train for which the signal is carried avail itself of any special orders which the train bearing such signals may receive, without a special order to that effect. When orders are duplicated to following trains, the understanding of each conductor and engineman must be separately written, and must be responded to by the person giving the order, as provided in the second preceding rule.[1]

In no case must the red signals be removed by operators until all trains have passed for which the order is intended.

Conductors and enginemen must look out for signals at telegraph stations.

Absence of proper signals at stations or on the road must be reported to the Superintendent.[2]

Under the system of moving trains by tele-

---

[1]. "A misunderstanding between trainmen and others interested in the running of trains is the most common cause of collision."—*Huntington.*

[2]. "Should a guard find any signal exhibited which ought not to be, or observe any other irregularity in the working of signals, or should he see any cattle or other obstruction on the line, or any defect in the signals, works, permanent way, or telegraph, he must report the same at the first station at which the train stops, and also on his journal."—*Eng. Standard.*

graphic orders, trains may be expected upon any part of the road at any time; this fact must be kept constantly in mind by employés.

In the event it is impossible or undesirable at any time to move trains by telegraphic orders, then and in that case trainmen will conform to the schedule and the rules and regulations governing the movement of trains incident thereto.

Trains must not pass a junction of two lines nor pass from a double track line to a single track line until the officials in charge have examined the register, kept by the agent or person in charge at such place, for the purpose of ascertaining whether trains due or past due have arrived, except in those cases where they have a special order from the Superintendent to proceed without stopping.

# CHAPTER VII.

### GENERAL INSTRUCTIONS TO CONDUCTORS.

The general direction and government of a train from the time it receives its passengers or freight until its arrival at its destination is vested in the conductor, and he will be held responsible for its safety and proper care; it is his duty to see that all rules and regulations and orders affecting his train are carried out.[1]

Conductors are responsible for the conduct of men employed upon the trains;[2] for the heating and ventilation of the cars, and for the signals, lamps, tools, etc., entrusted to their care. They must report defects in the air brakes (when such brakes are used) specifying the number of the car or engine on which it

---

1. "The duty of passenger, goods, cattle, mineral, and other guards consists in the general charge and management of the trains when they are moving on the line. They have general control over the enginemen, ordering them when to stop or to proceed at a different speed as they may deem right, or to shunt or move wagons or other vehicles."—*Gt. Nor. Ry., Eng.*

2. "When there are two guards with a train, the under guard must obey the orders of the head guard.

"Each train is under the control of the head guard, who must instruct the engine-driver as to the stopping, starting, and general working of the train. Whilst trains are within station limits, the guards are under the orders of the station master or person in charge."—*Eng. Standard.*

occurs. They will invariably require the air brakes to be tested, cylinders and connections examined, and also engine signal bell to be rung from the rear car of their train before leaving a terminal station.

They must report for duty, and in readiness to take charge of their trains, at least thirty minutes previous to the schedule time for starting, and as much earlier as they may be required, to assist in switching and making up of their trains.

They must be provided with a good reliable watch, which they must keep regulated by the standard clock of the company.

They must compare time with the engineman of the train before starting,[1] and know that he is provided with a schedule and a complete set of signals and tools.

They are also required to see that their trains are supplied with a full set of signals, and, when upon the road, they must see that such signals are used in accordance with the rules of the company.

Should a vehicle be attached to, or detached from, the rear of a train at an intermediate station, the conductor must see that the signals are removed to their proper places on the train.

---

1. " The guards are provided with timepieces. It is their duty to inform the engineman of the hour at every chief station or junction."—*Gt. Nor. Ry.*, *Eng.*

Each train (not including the engine) should be supplied with not less than twenty-four torpedoes,[1] five red lanterns, five red flags, five white lanterns, and five white flags, also with switch-rope, axes, saws, crowbars, chains, spare links and pins, buckets, oil, and such other tools and supplies as may be necessary for daily use, or in the event of accident or delay to the train.

If compelled by accident or other cause to move at an unusually slow rate of speed, or stop their train on the main track, they must take immediate action to signal any trains that may be approaching in either direction, as required by rule " L." They must keep in mind that nothing will justify a collision between trains, and that the prompt use of signals in the manner directed will, under all ordinary circumstances, prevent it.

Conductors and brakemen when meeting or passing other trains, or when approaching or passing a station, must be on the lookout for signals, and be prepared to do any thing which the expedition of business, or the safety of their train requires.

Conductors of trains must attend personally to all switches used by their engines or trains, and they will be held responsible for the proper adjustment of the switches used by them, ex-

---

1. " The earlier regulations governing the protection of trains have no reference to the use of torpedoes."

cept where regular switchmen are stationed. When there is more than one train to use a switch, conductors must not leave the switch open for the following train unless the conductor of such train is at the switch to take charge of it.

Conductors must see that street or public road-crossings are not obstructed by their trains while waiting. They will be particular when at junction stations, to see that no part of their train is allowed to stand on the crossings of other railways. This is especially important in regard to trains carrying passengers.

Conductors must enter the particulars of their trains in the register or arrival book at the ends of divisions. These books must be personally examined by conductors before leaving such stations, for the purpose of satisfying themselves as to the arrival of trains.

Upon double track roads, conductors must, before leaving the starting point, examine the register for the purpose of noting the departure of trains, their number and time of leaving.

They must visit the telegraph office before leaving terminal stations[1] to see if orders of any kind await them.[2] Conductors of freight

1. A terminal station is a station where a train is made up; upon a long line there will be several terminal stations; they are usually at the end of divisions or subdivisions.

2. "Every guard, before starting with his train, must examine the notices to see whether there is any thing requiring his

trains must at the same time report to the Superintendent the name of the engine and the number of empty and loaded cars in their trains. A similar report must be made by them upon their arrival at the end of their route.

Conductors must call the attention of the repairer of cars, or of the agent, in his absence, to any damage which may have been done to cars, or to any defects which may come to their knowledge, that the same may be repaired.[1]

Should complaint be made of the running of any car, the conductor must report it to the first car repairer, and enter the particulars on his report, giving the number and class of car; but if the conductor have reason to apprehend danger from such car before it can be inspected, he must have it detached from the train.

Conductors will report to the Superintendent any neglect on the part of car repairers to inspect each and every car that may pass such car repairers' stations; any neglect to carefully examine the running gear and brake fixtures of cars, and make such repairs as may be required;

special attention on those parts of the line over which he has to work, and he must, before going off duty, ascertain from the notices posted for his guidance, the time at which he is required to be on duty the following day."—*Eng. Standard.*

1. "It is an established rule, that if an accident happens to any foreign vehicle, the company on whose line it is running, is liable for the loss or damage, and also for all contingent loss or damage."—*Gr. Nor. Ry., Eng.*

any neglect to give special attention to passenger, baggage, mail and express cars. Repairers should not permit cars to leave their stations that are not in good running order. It is also the duty of car repairers to see that cars employed in the passenger service are properly washed, and that all the interior fixtures are kept clean and in good repair.[1]

When the wheel of a car or engine breaks, the conductor in charge must ascertain, by personal examination, the name of the manufacturer, and the date and number of the wheel. This information must be transmitted by letter to the Superintendent, and must also be noted in the train report.

In the event trainmen discover any defect or break in the telegraph, they must report the fact to the operator at the next station.

Conductors will advise the Superintendent of any dilatoriness or lack of attention upon the part of agents or others whose duties require their co-operation in the movement of trains.

Conductors, or their subordinates, must not, under any circumstances, undertake to carry or take charge of valuable packages, or make col-

---

1. "All plated reflectors in lamps are to be wiped with clean-washed leathers, kept solely for that purpose, and not rubbed with powder; when, however, they are much tarnished, they are to be cleaned with a little whitening."—*Gt. Nor. Ry., Eng.*

lections for individuals, unless authorized to do so by the proper officers of the company.[1]

Conductors will make a detailed report, in writing, to the Superintendent, of all accidents or injuries to persons or property, that may occur upon or in connection with their train, also the names of witnesses, if any.

They must report at the end of each trip the number of each and every car hauled in their train; the initials upon each car; the point from which taken; the place where left; whether loaded or empty; also, the class of car.

It is important that letters, way bills and dispatches should be delivered promptly.

When a trainman or other employé is returning to the station at which he resides, by a train other than that he is appointed to work, he must render all assistance in his power in the working of the train by which he travels, and obey any instructions received from the conductor in charge of such train.[2]

---

[1] "Conductors will not be concerned in any freight or express matter over the road by the passenger train, and will permit none to be taken by any person, except the agent of the express having contracts with the road, and will see that the express agents confine themselves strictly within the limits of their contract."—1854.

"Guards are forbidden to carry any description of package either for themselves, their friends, or the public, without proper authority in writing for the free transit thereof, or unless such package be properly entered on the way-bill."—*Eng. Standard.*

[2] "The guard must see that platelayers and other workmen of the company holding third-class passes, are kept as separate

## PASSENGER CONDUCTORS.[1]

Passenger conductors must make themselves generally acquainted with the duties of enginemen, baggagemen, brakemen, express messengers, mail agents, sleeping-car conductors, porters and news agents, and rigidly enforce the rules and regulations applicable to them upon their trains, reporting to the Superintendent any insubordination, neglect of duty, or misconduct upon the part of such men.

" When a deficiency of room occurs in a train while on the journey, guards must telegraph to the next station where carriages are kept, to as possible from the passengers. When a large number of workmen travel by the same train, carriages must be specially provided for their use, and they must ride in these carriages only."—*Eng. Standard.*

"All guards are to enter their time in the time-book every Friday or Saturday night at King's Cross; if this be not done, they will be liable each to a fine of 25 cents, and no money will be paid till the following week."—*Gt. Nor. Ry., Eng.*

1. "When there are more conductors than the number of trains running, those in waiting at either end of the road will be at the depots on the arrival and departure of all trains, as far as practicable, to aid in making up the departing trains, or discharging those arriving.

"They will see that extra cars are kept at the proper places upon the line for use in case of accident or other necessity.

"They will consider themselves to be, and act as, brakemen, when necessary."—1853.

"When on duty, conductors must be respectably dressed.

"Every man on passenger trains and at stations must appear on duty clean and neat."—1854.

"Every passenger guard must have with him his watch, whistle, and carriage key, and take in his van a red, a green, and a white flag, a box of detonators (not less than twelve), and a hand signal-lamp."—*G. W. Ry., Eng.*

have one or more in readiness to attach, on the arrival of the train."[1]

They must see that passengers are properly seated, and will not allow them to stand on the platforms of cars, while in motion, nor ride in the baggage, express, or mail cars, nor in any way to violate the rules and regulations of the company.[2]

They must be respectful and considerate in their intercourse with passengers, conveying politely any information desired, and in every way consistent with the rules of the company and the rights of others they will endeavor to contribute to the pleasure and comfort of passengers.[3]

They must collect a ticket or fare from each

---

[1]. Eng. Standard.

[2]. "They will attend to the wants of the passengers before the departure of their trains; to obtaining proper seats for ladies, and to a proper disposition of their baggage, and see that everything is done in a quiet but efficient manner, to ensure the departure of their trains at the appointed time.
"They will endeavor, with proper and gentlemanly discretion, to seat their passengers in such a manner as will best conduce to the comfort and safety of the whole."—1853.

[3]. "It is one of the special duties of the conductor to do any thing (not inconsistent with other duties or the regulations of the road) to accommodate passengers; to answer in a proper and civil manner all questions, and endeavor to leave a good impression on every one.
"They will make themselves acquainted with routes of travel in general, and especially those in the vicinity of the road, and on the great main lines in connection with the road.
"From their position they are able to exercise a material influence in turning the patronage of passengers to certain hotels along the line, as well as at the end of the road. The impropriety of this they will readily see."—1853.

passenger, and make reports of the same in the manner and form prescribed. Any passenger refusing to pay fare must be put off the train at the next station, but unnecessary violence must not be used in doing so.[1]

They must not permit drunken or disorderly persons to get upon their trains. It is their duty to maintain proper order, and they must not allow the vicious and unruly to indulge in rudeness or profanity.[2]

They must see that the doors of the carriages are properly closed, and in case of any unusual stoppage on the road, must request the passengers to keep their seats, except when necessary to alight.[3]

[1] "Guards, on arrival at a ticket-collecting station, must request the passengers to have their tickets ready, and must assist the ticket-collectors by opening and closing the carriage doors; they must not, however, collect or examine tickets, except under special instructions.

"The guard must take care that passengers enter the proper carriages for the places to which they are booked, and that no passenger is allowed to leave the train for the purpose of rebooking by the same train, with a view to evade payment of the proper fare."—*Eng. Standard.*

[2] "Carefulness is always earnestly enjoined, and the men on the train *must always act in concert.*"—1854.

[3] "In all cases of detention or stoppage, it is the duty of the guards to explain to passengers the cause thereof, and if there is no danger to them, to satisfy them of that fact, and endeavor to pacify those that may be annoyed.

"When a train overshoots a station, the guard is to order the engineman to put back to the platform, and not to allow the passengers to get out until the train has been stopped at the platform."—*Gt. Nor. Ry., Eng.*

"The guard must not allow any person to ride outside the carriages, nor must he permit any unauthorized person to ride

## Trains and Stations. 157

Smoking in the carriages, except in the cars specially set apart for that purpose, is strictly forbidden.

They will not permit beggars, peddlers, or gamblers to practice their vocations upon the trains.

They must notice the temperature of the coaches, and instruct brakemen about attending to the heating apparatus, and to the ventilators.

Shortly before reaching a station at which the train stops, they must pass through each car, except the sleeping car, and announce distinctly the name of the station they are approaching. They will require brakemen to repeat this announcement in each car, with the doors closed, twice in succession, as distinctly as possible, just before the train arrives at the station.[1] They will require brakemen to assist

in his van, or in any compartment or vehicle in which parcels or luggage may be placed.

"No carriage door must be opened to allow a passenger to alight from, or enter, a train before it has come to a stand, or after it has started."—*Eng. Standard.*

[1]. "The policeman, porter, or other person on duty at a station must, on the arrival of a train, walk the length of the train, and call out in a clear and audible voice the name of that station when *opposite the window of each carriage,* so as to make every passenger in the train aware of the name of the station; and particular care must be taken by the clerk in charge, policemen, and porters, to observe the indication of any passenger, that they desire to alight, by their knocking at the windows, or otherwise."—*Gt. Nor. Ry., Eng.*

ladies,[1] children, and infirm persons to get on and off the trains, and insist upon their being courteous to all persons. At terminal stations conductors will not leave their trains until passengers have alighted, and they must render them all needful assistance.[2]

They must not signal their trains to start while passengers are getting on or off the train.[3]

When the signal is made, the conductor should stand near the front end of the forward passenger car.[4]

They must report every instance of agents failing to give passengers an opportunity to procure tickets, reporting any neglect of an agent to open the ticket office of his station

[1] " When ladies are traveling alone, the guards are to pay every attention to their comfort ; and in placing them in the train, they must, if requested, endeavor to select a carriage for them (according to the class of their tickets) in which other ladies are traveling ; and if they wish to change carriages during the journey, the guards must enable them to do so."—*Eng. Standard.*

[2] " When they arrive at the end of their trip, they will not leave their passengers until the whole of the baggage is distributed, aiding in its distribution and generally attending to the wants of their passengers, especially ladies and infirm persons."—1853.

[3] " They will always bring their train to a dead stop to take up or leave passengers."—1853.

[4] " The signal for starting the train must be given by the guard blowing his whistle and showing a hand-signal."—*Eng. Standard.*

" The guards and other servants of the company must take their seats in the trains before they are in motion, so as to avoid the dangerous practice of jumping on the steps, or getting into the carriages after the trains leave the platform."— *Gt. Nor. Ry., Eng.*

before the arrival of trains, when the rules require it.

They must know that the cars in their trains have been inspected at terminal and other stations, as required.[1]

They are required to wear the prescribed uniform, and must never appear on duty without it.

Upon the arrival of a train at its destination, the cars in which passengers may have ridden must be searched by the conductor;[2] any articles found must be delivered to the agent at the terminal station. The articles must be sent by such agent to the general baggage agent, if not called for within forty-eight hours.

### FREIGHT CONDUCTORS.[3]

" The guard in charge of the train must satisfy himself before starting, and during the journey,

---

1. "They will also report all the interior defects of their cars, like the rattling of doors, windows, etc."—1864.

2. "Every first-class carriage is to be searched at the end of each journey by the head guard, and every second and third-class carriage by the second guard."—*Gt. Nor. Ry., Eng.*

"They will at all times render all the service in their power to forward the private business of the company (as well as its business for the public) in the hauling of wood and materials for use upon the road, and in bringing to the repair-shops cars and parts thereof which may be out of order and left upon any part of the line."—1853.

3. "Every head goods guard must have with him his watch, and whistle, a red, a green, and a white flag, a box of detonators (not less than twelve), a hand signal-lamp, a full set of tail and side-lamps, two or more spare coupling-chains, a brake-stick, two sprags, and two hand scotches."—*G. W. Ry., Eng.*

that the train is properly loaded, marshaled, coupled, lamped, greased, and sheeted; that the brakes are in good working order; and that the train is in a state of efficiency for traveling, and has the proper signals attached to it. He must also carefully examine the loading of any vehicles he may attach on the way, and if any vehicle becomes unsafe from the shifting or derangement of the load, must at once have the load readjusted, or the vehicle removed from the train.

"The guard must see that the chains on timber-trucks and on boiler-wagons are secured in order to prevent their getting loose whilst traveling. Foremen, guards, and shunters must take care that no timber-truck or boiler-wagon is allowed to leave a station or siding without the chains being first carefully examined and made perfectly secure and safe, and guards will be held responsible for seeing that they remain so during the journey.

"Before starting from a station, the guard must see that the wagons are properly greased, the coupling-chains and doors securely fastened, and carefully examine the loading and sheeting of the wagons, seeing that the goods are protected from rain and sparks from the engine; also, that no load is too high or wide, or in any way unsafe to travel. It is not sufficient for the guard, on commencing his journey, to see that all the wagons and their loads in his train are in a secure state for transit, but he must see that all these conditions are continued throughout the journey, especially with wagons that are taken on at intermediate stations, and those loaded with timber, cotton, wool, castings, machinery, and articles of great length and bulky construction."[1]

Freight conductors must make themselves generally acquainted with the duties of enginemen, firemen and brakemen, and enforce the rules and regulations applicable to them upon their trains, and report to the Superintendent any insubordination, neglect of duty, or misconduct.

They must see that the couplings, wheels, journals and brakes of the cars in their train are in good order before starting, and will inspect them, when their duties will permit, or as often as the train stops to take water or arrives at a meeting or passing point.[2]

They must station the brakemen at their

---

[1]. Clearing House Standard, England.

In England the great bulk of the freight traffic is transported upon open or flat cars, the property being protected by sheeting or tarpaulin. The water-tight box car so universally used in this country is practically unknown there.

[2]. "They will frequently examine the cars of their trains to see that all nuts and screws are up to their bearings, and the cars in order; that they are properly oiled—not oiling

respective posts[1] on the train and see that they keep their position and use the brakes with discretion and good judgment — particularly when descending heavy grades.[2]

They are positively forbidden to take any loaded cars into their trains without a way-bill.

them at random, but when needed; and for this purpose will see that their trains are supplied with such tools as may be wanted, as well as oil for the bearings.

"They will not allow repairers to attach their repair cars to their trains, unless it shall be necessary in order to forward some very urgent piece of work."—1853.

"They must examine carefully and minutely every wagon, whether loaded or not, and its covering, the axle-boxes, the fastenings of its doors and side-flaps, etc., etc., and the way in which the goods are placed in the trucks, so that large loads may not overhang, or be too high; they must compare the road-bills with the wagons; see that they are placed in the proper position in the train; that they are entered correctly and properly labelled."—*Gr. Nor. Ry., Eng.*

1. "No goods, cattle or coal train, may start without one brake-van at the least, which must be placed behind the train; and, in case of two brake-vans in one train, one of the guards must ride in each, so as to work both the brakes."—*Gr. Nor. Ry., Eng.*

In England the style of the car used prevents brakemen from traveling backwards and forwards upon the top of the train, as in this country.

"The freight conductors must ride on the tender facing train, or else on the rear car."—1854.

"The guard must ride in his brake-van, and not upon the engine or in any part of the train; he is forbidden to pass over the tops of the carriages" (passenger cars) "when in motion."—*Eng. Standard.*

"They will consider themselves to be, and act as, brakemen when their train is in motion."—1853.

2. "A rear brakeman, by leaving his post for a short time to have a friendly chat with his next brakeman, has been the immediate cause of such mischief" (*i. e.*, the cause of a collision).—*The Trainmasters' Assistant*, p. 124.

If way-bills are not ready, they will not take the goods, but report the fact at once to the Superintendent, giving the name of the agent.

They will also refuse to take cars that, in their judgment, are unsafely loaded, reporting the reason therefor to the Superintendent.

They must not move cars from an intermediate siding or private switch without way-bills have been furnished them by the agent at the last station they pass before reaching such siding or switch ; or in the event they do take freight from such places without bills, they must report the same at the first station where there is an agent, when a way-bill must be made, at the prices named in the tariff, from the place where the freight was taken.

Passengers should only be allowed upon such freight trains as the General Manager may elect, and not upon such trains unless they are provided with tickets.[1]

They must know that the cars contained in their train and reported as being empty are so in fact.

They must see that the cars are always locked, except when loading or unloading freight.

---

1. "They are prohibited from allowing persons to ride upon the freight cars in the train ; nor will they allow them to ride in any passenger car that may be attached to their train without paying for it, even though they are in the employment of the company, unless they have a written free pass from a proper person ; except in cases of accident to the road or trains, when they will act as the interest of the company may, in their judgment, require."—1853.

They must also see that the windows of cars are fastened.

They must take loaded cars from all stations when they can haul them, although their train may be behind time.[1]

They must carefully note (check) upon the way-bill each article left at a point where the company has no agent, attaching their signature to the bill. If any goods are damaged or missing, they must make a note of the same upon the face of the way-bill.

When loaded cars, destined for any station, are left at another station, the way-bills must be left with them.

They are required to treat those in charge of stock politely, and render them every assistance possible in taking proper care of their stock.[2]

[1] "The object of running freight trains being to do the business of the road, and not altogether to make time."—*Western Road*.

"In passing over the road, they will attach to their trains all the loaded cars which may be ready for them, in the order in which they come, whether at regular stations or side-tracks, till they have a full train; but a loaded car is not deemed ready for the train until the agent has the doors locked and fastened, and a way-bill ready; and the conductor will call for a way-bill in all cases, that he may be sure of the proper distribution of all the cars or freight in his train. They will take all empty cars from side tracks where they are not wanted, and draw them where they are required, if in the direction in which they are running."—1853.

[2] "They will see that all live stock upon their trains are fed and well taken care of, and put the cost of feeding them upon the proper way-bill, for collection by the station agent."—1853.

They must not permit persons in charge of live stock to ride free upon their trains without a written permit from the proper official.

Freight conductors will be held responsible for freight while in their charge.[1]

In leaving loaded cars at a station, they will leave them at the most convenient place for unloading, and in cases of this kind they must act in harmony with the agent.

They must personally check from the car the way-freight unloaded and delivered from their trains; the property must be checked in the presence of the agent, and in the event there is any freight over, or short, or damaged, the facts must be noted on the way-bill.

In loading and unloading way-freight, they must be particular to see that the property contained in the car is in a safe position, so as not to be afterwards affected or damaged by the oscillation or jar of the train.

They must see that care is exercised in loading and unloading way-freight, using every possible effort to prevent loss or damage.

They are required to deliver way-freight on the platform at the freight house, or at such other proper and reasonable place as the agent may designate.

[1] "They will be held accountable for any loss or damage to freight caused by rough handling, by carrying it past its destination, by wrong delivery, or by neglecting to take it at way-stations, when requested to do so by station agents."—1854.

In delivering way-freight the train must not be delayed longer than necessary.[1]

Freight trains must stop at the places specified in the schedule, unless, on approaching a station or siding, a signal is given by the agent or signalman that it is not necessary for the train to stop. When this is done, the train may run past the station or siding without stopping, unless there are cars or goods to leave, when the engineman will have instructions from the conductor to stop. In the case of a train timed to stop at a station or siding " when required," the engineman of such train must stop at the station, or siding, unless he receives a signal to proceed without stopping.[2]

They must not permit persons not duly authorized to enter cars or handle freight on their train.

They will report any confusion or want of method upon the part of agents in loading freight.

[1] " They, with the brakemen, will render all aid in their power, on the arrival of their train at a station to enable them to leave in the shortest space of time; that as much time may be used in running, and as little in stops as possible."—1853.

[2] " In order to prevent the unnecessary stoppage of the train, if the engine has a full load, and can not take more wagons on, and has nothing to leave at the station, the guard must give a green signal, to indicate to the clerk in charge that he has his full load, and can not take more. And it will be the duty of the clerk in charge to count the wagons in the train signaled as fully loaded, in order that inquiry may be made, in case of any improper refusal to stop on the part of the engineman."—*Gt. Nor. Ry., Eng.*

They must use great care in the handling and loading of coal oil, and under no circumstances will it be loaded with other freight that can be damaged by it. So far as practicable, it should be loaded in stock cars.

When waiting upon sidings, and at other times, they must exercise great watchfulness to prevent cars from being broken open, and the contents thereof stolen or damaged.

Should a loaded car become disabled, or from any cause be left on a side track, where there is no agent, the conductor will deliver the way-bill to the agent at the next regular station, and endorse on the way-bill when, where, and why left, and report the facts to the Superintendent.[1]

When it is necessary for trainmen to transfer freight from one car to another, the reason for such transfer must be noted on the face of

---

[1] "When it becomes necessary to switch off cars on account of accidents, storms, or from lack of power, freight conductors must examine the contents of such cars, and if they contain perishable property, it must be re-loaded by them and forwarded immediately.

"When, from a train being overloaded, or for other cause, it becomes necessary to leave where they do not belong, any loaded cars at any other side track than at a regular station, they shall leave a man from their train with them, till a train passes which can take them to their destination. They will also note upon the proper way-bill what cars they have left, and where they leave them; and when it becomes necessary to leave any disabled loaded cars at any other than a regular station, if they have not empty or partly loaded cars in their trains to transfer the loading into, they will leave a man with them until empty cars arrive, and the loading is transferred and started for its destination."—1853.

the way-bill, and the number of the car into which the freight was transferred must be inserted, and the number originally entered crossed out.

Conductors, when at stations doing business, will attend personally to the switching.

They must not absent themselves from duty without permission from the Superintendent.[1]

While waiting at stations, conductors of freight trains will do such switching as may be reasonably required by the agent.[2]

They must be sure that no cars have become detached from their train and left on the main track, and when cars are left on a siding, they must see that the brakes upon such cars are securely applied, and the wheels carefully blocked, to prevent such cars from being moved, or interfering in any way with trains or cars upon other tracks.

They must indorse their names, in the place provided, on the back of each way-bill carried by them.

They must make immediate and complete reports to the Superintendent of all unusual detentions to their trains, and in case of accident

---

1. "Goods guards must not leave their trains until they have been delivered over to the foreman, yardman or shunter."— *Eng. Standard.*

2. "They, with the brakemen, when not otherwise employed, will render what aid they can in wooding and watering, to shorten their stops."—1853.

to cars, resulting in damage or loss of property, will at once telegraph or write all the facts to the Superintendent.

They must carefully enter upon their reports the number of cars taken from and left at each station, as already described, and make such other regular returns as may be required of them.

### GENERAL INSTRUCTIONS TO BRAKEMEN.

While on duty, brakemen are under the direction of the conductor.[1]

They are charged with the immediate management of the brakes, the proper display and use of signals, and the lights, stoves, water and gas-fixtures of trains.

They will be furnished, upon the requisition of the conductor, with a full set of train signals, which they must keep in good order, and at hand, ready for immediate use.

The rear car of every train must be a brake-car. A man must always be on the rear car of trains; provided, however, that when stopping a train he may set the brakes upon forward cars after having set the brake on rear car.

In the absence of automatic brakes, they are required to stop their trains at stations, and

---

1. "At stations, it is their duty to assist in taking on wood and water, and, when not on running duty, must assist at the station in whatever work may be required of them."—1853.

control them when descending heavy grades without the whistle signal of the engineman.[1] In damp or frosty weather, the brakes must be applied sooner than usual to prevent running past the station.

Brakemen must obey the order to apply brakes instantly without waiting to ascertain the occasion of the signal.

The post of the rear brakeman (or flagman) is on the last car in the train ; he must not leave his post while the train is in motion except to protect it ; he must be provided with the necessary signals, and must see that they are displayed at the rear of the train, in accordance with the rules ; and in case of detention or accident to the train, he must immediately go back, as directed in such rules, for the protection of trains; he must do this promptly and without waiting for a signal from the engineman or instructions from the conductor.

[1]. " They are not allowed to slip the wheels only in cases of danger, and never upon the ordinary occasion of stopping at a station ; observing strictly when the engineman shuts off steam on approaching a station that it is a signal, without waiting for the sound of the whistle to apply the brakes, using judgment in order to stop at the proper place at the station without allowing the train to press hard upon the tender or engine, allowing the engineman to stop the engine and tender without causing them to draw or press upon the train."—1853.

" In traveling down steep inclines, guards must, in order to steady the trains and assist the engine-drivers, apply the rear brake, care being taken not to skid the wheels except when a train is approaching at too great a speed a station at which it is timed to stop, or when the brakes are specially whistled for by the engine-drivers."—*Eng. Standard.*

The front brakeman is charged with the duty of protecting the train with signals, when, from any cause, the fireman is unable to perform this duty.

In case the train parts on the road, the rear brakeman must immediately apply the brakes and stop the cars, and then send forward the most reliable person he can command to make danger signals, while he protects the detached portion until the engine or front part of the train returns.

When an assistant engine is attached to the rear of a train, it must be considered as a part of the train, and in case of accident or detention, the brakeman must go back as in other cases.

They must examine the running gear of the cars at the various stopping-places, reporting to the conductor any defects they may observe.

Brakemen will be held responsible for the brakes and the condition of the coupling apparatus. It is their duty to see that these are in good order before trains start.

It is the duty of an employé who opens a switch upon the main track to see personally that it is afterwards closed and locked.

### PASSENGER BRAKEMEN.[1]

Passenger brakemen will report for duty at the time appointed, open the doors of the cars,

---

[1] "Brakemen on passenger trains will be required to wear coats or overalls when on duty."—1854.

and assist conductors in the proper disposition of passengers, and will aid them in all things requisite to the prompt and safe movement of the trains, and the comfort and convenience of passengers.

They must give special attention to the proper heating and ventilating of the cars, keeping a moderate, uniform temperature, and see that the air does not become impure.[1]

At all stopping places they will twice distinctly announce the name of the station, and the length of the time the train will stop, when such stoppage exceeds two minutes.

They must assist the conductor in preserving order, and will not permit passengers to stand upon the platforms while the train is in motion, nor in any way to violate the rules of the company.

They must be respectful to passengers, and give polite attention to their wishes, avoiding, however, any unnecessary conversation.

When it is necessary to pass through sleeping cars, they will do so quietly so as to avoid disturbing passengers.

When not otherwise engaged, they will stand at the door of the car, ready to respond to the

---

[1]. " The brakemen must light the car lamps, and make and keep up the fires in the cars, fill the water-casks or jars, and do such other work on the train as the conductors require.

" He will see that the water-casks are filled with clean water and in warm weather that they are well iced."—1854.

signal of the engineman, and they must occupy this position whether the train is equipped with automatic brakes or not.

They are required to see that the water closets of cars are kept in a cleanly condition.[1]

#### FREIGHT BRAKEMEN.

Freight brakeman will report for duty at the time appointed by the dispatcher, and will assist the conductor in the switching and making up of trains.

They must not leave their posts while the train is in motion, nor take any other position on the train than that assigned to them by the conductor.

They must assist in loading and unloading freight.

They are required to stop their trains at stations, and control them when descending

---

[1]. "There is no water closet in the train, no passage through the cars, and no means of communicating with the conductors. Robberies have often been committed in the carriages. Each compartment is lighted at night by a lamp in the roof and warmed in winter by flat tubes of metal filled with hot water and placed under the feet of the passengers on the floor."—*Private letter from Rome describing passenger cars in Italy.*

"The guards must immediately open the door of any carriage from which passengers may require to alight for the purposes of nature, etc., particularly at those stations where the engines take in water. The guards must, on all occasions, represent to passengers the necessity for their resuming their seats quickly for the prevention of delay, and they must avoid all loss of time on the journey."—*Regulations English Road.*

"Guards, porters, policemen, etc., etc., are forbidden to use the water closets provided for the public, and will be fined for so doing."—*Gt. Nor. Ry., Eng.*

heavy grades, without the whistle signal of the engineman. The brakes must not be applied so as to slip the wheels, and on heavy grades the brakes should be changed frequently from car to car so as to avoid heating the wheels.

#### TRAIN AND STATION BAGGAGEMEN.

The duties and responsibilities of these officials are explained in a separate volume in connection with the business of the baggage department and the traffic incident to it.[1]

#### ENGINEMEN.

When passing over the road without a conductor, they will be held responsible for the faithful and intelligent use of all the precautions required by the rules and regulations governing the movement of trains. They must, therefore, familiarize themselves with such rules and regulations, including those for the government of trainmen.

They are intrusted with the lives of passengers and the property of the company, including that which is intrusted to it for transportation. It is important, therefore, that they should not only attend promptly to the signals given them, but they should be vigilant and cautious, not trusting blindly to the signals they may receive, nor the rules and regulations provided for their safety.

---

1. "Baggage Car Traffic," by Marshall M. Kirkman.

The engineman, before commencing his day's work, must examine the notices posted for his guidance, in order to ascertain if there is any thing requiring his special attention on those parts of the line over which he has to work.

"The engineman must stand up and keep a good look-out all the time the engine is in motion, and the fireman must also do so, when he is not necessarily otherwise engaged."[1]

It is the duty of enginemen and firemen at all times to keep a sharp look-out to see that no portion of the train becomes detached without their instantly observing it.[2]

Enginemen are under the direction of conductors when upon the road, in all things not in conflict with established rules and regulations.

Enginemen will observe the orders of the inspectors and master-mechanics in regard to the working of their engines and the proper use of fuel and stores.

They must obey the orders of the yard-master or person in charge in regard to switching and making up trains.[3]

---

1. Eng. Standard.

2. "With the firemen, they will often alternately look around to see that all is right with the train while passing over the road, or standing with their train at stations, and to attend to signals from the conductor, for starting forward or backward."—1853.

3. "The engine-driver must afford such assistance with his engine as may be required for the formation, arrangement, and dispatch of his train."—*Eng. Standard.*

They must not start their trains till directed by the conductor, nor till the bell of the engine has been rung. They must start with care, and it is their duty to see before they get beyond the limits of the station that no portion of their train is detached.[1]

They are required to start and stop the train slowly, otherwise the couplings and chains are liable to be broken.

In stopping their trains, they must pay particular attention to the state of the weather and the condition of the rails, as well as to the length of the train. These circumstances must have due weight in determining when to shut off the steam. Terminal stations must be entered with special care.

They must be careful not to shut off steam suddenly (except in case of danger), so as to cause a concussion of the cars, by which trainmen and others may be injured. Enginemen of stock trains are required to be particularly careful in starting and stopping their trains.

[1] "When a passenger train is about to start from the station or ticket-platform, the signal to start given by the guard merely indicates that the station duty or the collection of tickets is completed ; and previous to starting the train, the engine-driver must satisfy himself that the line before him is clear, either by observation, or by obtaining, by means of his whistle, the exhibition of the necessary signal, as the circumstances of the case may require, and, when starting, the fireman must look back on the platform side until the last vehicle has drawn clear of the platform, to see that the whole of the train is following in a safe and proper manner, and to receive any signal from the station-master or guard that may be necessary."
—*Eng. Standard.*

They must know exactly what time is allotted them in the schedule, and they must not start from a station, even though they receive a signal from the conductor, unless they can reach the next station without encroaching upon the rights of other trains.

They must have their engines in good working order, supplied with the necessary signals, stores, tools,[1] fuel and water, and the steam up ready to attach to the train at least thirty minutes before the schedule time for starting, and as much earlier as directed by the proper official.

They must see before leaving the engine house that the spark-arrester and wire netting over the smoke-pipe and the ash-pan of the engine as well, are all in good condition.[2]

In running passenger trains, enginemen must observe great care in the manner of working the automatic brake. It must be applied when the engine is first attached to the train, before

1. Each engine must be supplied with twelve torpedoes, two white lamps, two white flags, four red lamps, four red flags, two green flags, two green lanterns, one yellow flag, one yellow lantern; also with a pair of screw-jacks, extra spring-hangers, and such other tools as may be necessary to operate the engine or provide for accidents or delays.

2. "They will be particular to see that the chimney is kept in order, so as not to throw fire. They will not empty their sparks between the extreme switches at any station, unless a proper place be provided for them. Where no place is provided, select the most suitable beyond the switches, putting them down an embankment, if possible, so as not to disfigure the line."—1853.

starting from the station, to make sure that it is in working order; in making regular stops, it must be applied in such manner as to avoid injury to the brakes, or discomfort to the passengers. Especial care should be taken with short trains to apply the brake sufficiently early to obviate this difficulty.[1]

The brakes must not be relied upon when approaching railroad crossings or other hazardous points, but steam must be shut off, and the train, whether passenger or freight, held under such control as to prevent running past the objective point before stopping.

Enginemen are required to see that the bell-cord is not obstructed by fuel or otherwise. It must not be unfastened until the end of the trip, and when more than one engine is attached to the train, the bell-cord must be attached to the leading engine.[2]

---

1. "Should a passenger train, in stopping at a station, overrun, or stop short of the platform, the engine-driver must not move the train back or draw it forward until he receives instructions from the guard in charge to do so. Station-masters, guards, and others, must at once take steps to prevent passengers leaving the carriages that are not at the platform; and as soon as the guard in charge has satisfied himself that all carriage doors are closed, and that no passengers are entering or leaving the train, he must instruct the engine-driver to put back or draw up to the platform, as may be required. The engine-driver must sound his whistle before moving his train."
—*English Standard.*

2. "When two engines are employed in drawing the same train, the engine-driver and fireman of the leading engine are responsible for the observance of signals; the engine-driver of the second engine must watch for, and take his signals from

"Wagons must not be shunted into sidings, nor to other wagons upon the main lines, without remaining attached to the engine, except the wagons are attended by a brakeman or other person prepared to put down the wagon-brakes or apply sprags, as the case may be, so as to prevent their coming into violent contact with other wagons or vehicles, or fouling other lines. When wagons require to be shunted into incline sidings, the trucks to be moved at one shunt must be limited to such a number as the engine can push up without going at a violent or excessive speed."[1]

They must promptly obey all signals given, even though they may think such signals unnecessary. When in doubt as to the meaning of a signal, they must stop and ascertain the cause, and, if a wrong signal is shown, it is their duty to report the fact to the Superintendent.[2]

They must notice whether watchmen and

the engine-driver of the leading engine, but the engine-driver of the second engine is not relieved from the due observance of all signals regulating the safe working of the line. Great caution must be used in starting such a train to prevent the breaking of couplings."—*Eng. Standard.*

1. Eng. Standard.

2. "The engine-driver and fireman must pay immediate attention to and obey all signals, whether the cause of the signal being shown is known to them or not. The engine-driver must not, however, trust entirely to signals, but on all occasions be vigilant and cautious. He must also obey the instructions of the officers in charge of stations."—*Eng. Standard.*

flagmen are at their posts, and report to the Superintendent any neglect of duty that they may observe.

They must also report the absence of lights at switches, where such lights should be shown.[1]

They must approach and pass stations where their trains do not stop with great caution.

When trains are running over the road in heavy storms, or immediately after such storms, enginemen will run very cautiously, and without regard to making schedule time. They should run slowly and cautiously in approaching curves and places where the track is likely to be washed away.

Great care should be taken to prevent the killing of live stock, and engines must come to a full stop, if necessary, to avoid killing or injuring stock.[2] Conductors and enginemen must report to the Superintendent, in writing, giving full particulars whenever stock is killed or injured by their engines or trains.

"Engine-drivers, after taking water from tanks or water-columns, must be careful to leave the hose or water-crane clear of the main line and properly secured."[3]

---

1. "The absence of a signal at a place where a signal is ordinarily shown, or a signal imperfectly exhibited, must be considered a danger signal, and treated accordingly, and the fact reported to the signalman or station-master." — *Great Western Ry., England.*

2. "Pass all roads cautiously; be careful not to frighten horses, and at Blank take extra care."—1853.

3. English Standard.

They must not permit burning cotton, waste or hot cinders to be thrown from the engine or tender while in motion, and must use every precaution when passing bridges, culverts, buildings, and wood-piles to prevent the same from taking fire from their engines. The dampers of ash-pans must be closed when passing over wooden bridges or culverts.[1]

They will not be permitted to clean or empty their ash-pans on the main track, except at points designated by the Superintendent.

Enginemen must so arrange their fires as to avoid any unnecessary emission of smoke from their engines while standing at or passing stations.

They must not leave their engine during the trip, except in cases of necessity, and, when absent from it, they must leave the fireman or some other competent person in charge.

The store-keepers will provide them with tickets for wood and coal, and they will not be furnished such supplies except upon the delivery of a ticket to the person in charge, for the correct amount supplied them. It is important, therefore, that they, at all times, keep a sufficient supply of tickets on hand.

When a conductor is disabled, the engineman will have full charge of, and be held responsible for the safety of the train until

---

[1] "Pass all important bridges carefully and at a reduced speed, with the ash-pan closed."—1853.

an authorized person assumes charge of it. Enginemen must never leave their engine when steam is up without shutting the regulator, throwing the engine out of gear, and applying the tender-brakes.

They must report the condition of their engines to the inspector or master-mechanic at the end of each trip.

They will, at all times, assist, when called upon, in making any repairs to their engine that may be necessary. When required to work in the shops, they will be subject to shop rules and regulations.[1]

When enginemen or firemen become morose and sour from long service, they should be retired.

### FIREMEN.

Firemen, when on duty upon the road, are under the direction of the enginemen.[2]

They will obey the orders of the master mechanic or inspectors of engines in regard to the use of fuel and the proper manner of firing.

They must be on their engines at least thirty minutes before time of starting, and conform to any directions they may receive from the enginemen.[3]

[1]. "When not on running duty, they will assist in the machine shop, and conform to its rules."—1853.

[2]. "They will act under the direction of the engineman, and will aid in the small daily repairs and cleaning of the machine."—1853.

[3]. "They must see that the boilers are properly filled before firing up; that the fires are kindled in proper time, and

They must supply the engine regularly with fuel and water, at the discretion of the engineman. They must ring the bell when required, and must assist in oiling, and apply the tender-brake, in accordance with the orders and signals of the engineman.[1]

They will assist in keeping a constant lookout upon the track, and must give the engineman prompt notice of any obstruction they may perceive.

They must make themselves familiar with train rules, including those that apply to the protection of trains, and must understand the use of signals, and be prepared to use them or respond to them promptly and discreetly.

They must take charge of the engine, should the engineman at any time be absent, and will not leave it until his return, nor suffer any unauthorized person to be upon it.

They will not attempt to run an engine in the absence of the engineman without permission from the master mechanic, unless they are directed to do so by the conductor or other

that all the working-joints of the engine are kept well oiled, together with such other duty as the enginemen may require of them."—1854.

"They are strictly forbidden to throw fire or sticks of wood upon the road, as also to interfere in any manner with the running of the machine."—1853.

1. Before arriving at the station where they are to take wood, they will pile up their remaining wood in the front part of the tender, that the wood from the station may be taken in with the greatest dispatch.

authorized officer in consequence of some special emergency.

They must keep their engines clean,[1] and must assist when not otherwise engaged in making such repairs as may be required.

When at work in shops, they will be subject to the rules and regulations governing shop labor.

### INSPECTORS OF ENGINES.

Inspectors of engines will obey all orders of the master mechanic, and must report to him.

They are required to ride upon the engines and instruct enginemen and firemen in regard to the proper working and firing of engines, so as to obtain the best results in the consumption of fuel and stores.

They must study the capacity of the various engines.

It is their duty to see that the regulation pressure of steam is not exceeded, and that the boilers are washed as often as necessary.

They must see that engines are equipped with signals, tools, and articles necessary to their efficient working, and that injectors, pumps, brakes, and other fixtures are in good working order.

They will advise the Superintendent of the number of cars to be allotted to each class of

---

[1]. "During the passage, whenever they have an opportunity, they will wipe the connecting-rods and most exposed parts of their machine, keeping it as clean and neat as possible."—1853.

engines, and report to him when engines of through freight trains are not given cars to their full capacity, or when an engine is overloaded.

They will consult with the shop foremen in regard to the daily condition and requirements of the engines running upon their divisions.

They will report to the master mechanic and Superintendent the qualifications of enginemen and firemen, and any violation of rules or neglect of duty which may come to their knowledge, and keep them advised of all matters relating to the economical and efficient working of the engines and their crews.

### YARD MASTERS.

They will have charge of the yard and sidings at stations where trains are formed, the movement of trains in connection therewith, and of the yard force employed thereat.

When the business is not sufficient to require a yard master, the duties of the office, generally, will be performed by the agent.

They are responsible for the dispatch of trains, the prompt movement of cars within the limits of the yard, and the proper position of switches.

They must carry out the orders of the Superintendent in regard to the distribution of cars, the making up of trains, and assigning motive power therefor.[1]

They must give directions for switching and placing cars in proper position in trains, and see that such trains leave on time.[2]

They must see that the train force is ready for duty at the time required, and that both enginemen and conductors are supplied with schedules, signals, lamps, tools, and such fixtures as are required for the safety and good management of trains.

They must not permit a train to start with an engineman, conductor, or brakeman who is unfit for duty, nor fail to report such an occurrence to the Superintendent.

They must see that the yard is kept in good order, and that cars requiring serious repairs are sent to the shop.

It is their duty to see that car repairmen perform their duties of oiling, cleaning, inspecting, and repairing cars in a thorough and efficient

---

[1]. "At any terminus, or large station where carriages are kept, the station-inspectors are to see that they are always in good order, and, before being formed into a train, that every carriage or other vehicle has its proper supply of roof-lamps trimmed; that it is cleaned inside and out, and the glasses and handles made bright. They are also to see to the screwing up of the connections, and that the buffers of the several carriages forming the train press against each other, and recede about an inch when screwed up, and also to take care the doors on the off-side of all carriages are locked."—*Gt. Nor. Ry.*, *England.*

[2]. The duties of yard masters referred to herein, refer more directly to freight trains.

manner. Any neglect they may observe must be reported to the Superintendent.[1]

They must see that a record is kept of the number of each car, the date it arrived and departed, and that daily telegraphic returns of the same are made.

[1] "At stations where carriage-examiners are kept, the station master, or person in charge, must, before starting the train, satisfy himself that the examination of it has been completed, and that, so far as the carriage-examiner is concerned, the train is all right and fit to proceed. At stations where examiners are not kept, steps must be taken to remedy any defect that may be observed in the running of the vehicles, by supplying oil or grease to the axle-boxes of any that may require it, or removing the defective vehicles from the train, as may be found necessary."—*Eng. Standard.*

## CHAPTER VIII.

#### TELEGRAPH OPERATORS.

Telegraph operators at stations will observe the wishes of agents, when such observance does not interfere with their duties as operators.

They are required to be on duty without intermission during business hours, and must not leave their offices without permission from the Telegraph Superintendent.

They must not leave their post until relieved. The operator going off duty must advise the operator coming on in regard to unfinished business and the position and character of trains upon the line.

Offices will be in charge of the day operator.

Where two or more day or night operators are employed, they must not all be absent at their meals at the same time.

Operators at way-stations must be in their offices on Sundays twenty minutes before each train is due, and remain in the office until the train is reported as having passed the next telegraph station.

Operators must not leave their offices while a train is at the station unless the business of such train requires it.

They must be courteous in their intercourse with each other and with all persons transacting business at their offices.

Night operators must report to the home office every half hour from 9 P. M. till 7:30 A. M.

At one minute before eight o'clock A.M. each day, excepting Sundays, all business must be suspended, for the purpose of enabling the home office to report the exact time, and operators and others on the line must forthwith regulate their timepieces to correspond with such report.

At nine o'clock in the morning of each day except Sunday, all business will be suspended, for the purpose of sending car reports to the home office. In sending these reports, care must be taken to punctuate them properly.

All orders and instructions must be carefully preserved and filed for purposes of reference.

When there is a delay of more than fifteen minutes in sending a message, the particulars of the delay must be noted on the back of such message.

When practicable, messages received for transmission should be read aloud before being sent, either by or in the presence of the sender.

They will be held responsible for the prompt delivery of messages at their stations.

They must exert themselves to obtain answers to message promptly when answers are required.

In case parties to whom messages are addressed can not be found, the office at which the message originated should at once be notified.

When answers are required to messages and are not forthcoming, the reason should be explained as soon as practicable.

They must retain copies of all messages sent and received, also copies of reports of trains.

They must consider telegraphic messages as confidential in their nature, and they must not permit them to be read, except by those to whom they are addressed, nor will they make them the subject of conversation or remark.

Passes received by telegraph must be written with ink, and must contain the name of the office where received, the date and time of receipt, including the signature of the operator.

In transmitting messages, the circuit must be firmly connected, the writing must be plain and legible, and care must be exercised to punctuate in accordance with the communication itself.

In case of interruption or trouble to the line, operators must make diligent inquiry as to its location. The facts must at once be communicated by signal or otherwise to repairers or to trackmen, diligent efforts being made by the operators themselves to remedy the break.

Care must be exercised to protect instruments from being injured by atmospheric electricity.

Instruments must not be taken apart, but must be carefully preserved in good order, and none must be kept on hand that are not in use. Instruments or fixtures not in use or in a damaged condition must be returned to the home office.

The telegraph must not be used for the transmission of communications which may be sent by train without detriment to the interests of the company.

They must promptly report the departure of each train to the Superintendent; the arrival of trains must also be reported at terminal stations.

Conductors are instructed to report to the Superintendent when they are over fifteen minutes late; in the event they neglect to do this, operators must inquire as to the cause of the delay, and forthwith transmit to the Superintendent the result of these inquiries, also the name of the conductor and the number of his train. If the delay was caused by a hot journal, the number of the car or engine upon which it was located must also be reported.

They must see that they are supplied with proper signals for stopping trains, and will have them convenient and in order for immediate use when occasion requires. They must see that their signal lamps are properly trimmed and filled before dark each day.

They will observe the rear of all trains passing their stations, and if red signals are not displayed, they will at once report the omission to the Superintendent.

Particular attention must be given to the adjustment of relays when trains are behind time, or when the current is weak.

They will not be allowed to undertake to teach students how to telegraph without permission from the Telegraph Superintendent.

Conversation of a personal nature between operators must not be allowed to interrupt business.

Improper language or profanity should not be indulged in on the line.

Quarreling and contention amongst operators for the use of the circuit is reprehensible in the extreme. Should the current be interrupted while an operator is using the circuit, he should stop and ascertain the cause; should the interruption be occasioned by another operator having business entitled to preference, in accordance with the rules, he will give way to such operator; but in the event this is not the case, will signal such operator, "Close your key; you are breaking," closing his own key immediately thereafter. If the signal is not at once complied with, the operator will permit his key to remain closed until he can proceed without interruption, when he will at once report the case to the Telegraph Superintendent.

They must disconnect their instruments from the circuit when they leave their offices.

Offices at which there are night operators must be kept open continuously. Other offices must be kept open from 7:30 A.M. to 8 P.M.

#### TELEGRAPH REPAIRERS.

Telegraph repairers must pass over the road frequently.

They must closely observe the condition of the line, making a careful examination of the connections at the various offices.

They will report to the Telegraph Superintendent each morning the part of the road they propose visiting during the day.

When traveling upon the road, they must ride in the rear end of the last car, so that their view of the line may be unobstructed.

They must keep the telegraph poles in proper position, the wires connected, insulated and clear of all obstructions, and must make necessary repairs, calling upon the foremen of sections when assistance is required.

As they proceed, they must ascertain at the various telegraph stations how the line is working.

When upon duty, they must carry with them

the tools required in their business, such as pulleys, vises, plyers and file, hooks or cleats, insulators, etc.

They must see that operators and section foremen are supplied with wire, insulators, and other fixtures required in making repairs.

It is their duty to instruct operators and foremen of sections in reference to splicing wire and making other repairs necessary from time to time.

In case of a break or obstruction to the line, they must make diligent search for its whereabouts, and, having ascertained its location, proceed at once to make the necessary repairs. Having done this, they will report to the Telegraph Superintendent the location of the difficulty and its cause.

## CHAPTER IX.

### AGENTS.[1]

The duties of agents in connection with the baggage department and its affairs are treated of separately in connection with the baggage department and its traffic, as already explained herein.

#### RULES REFERRING TO THE PASSENGER TRAFFIC.

Agents must be careful to keep on hand, at all times, a supply of tickets sufficient to answer the wants of the business of their stations.

Agents must use every exertion to supply passengers with tickets before such passengers enter the cars, but they must not sell tickets to stations at which the train does not stop.

Agents must not sell tickets to persons who are unfit to take care of themselves, or who might endanger the lives of passengers, or prove an annoyance to them.

---

1. The rules and regulations governing the collections and accounts for freight and passenger traffic (Revenue) and the Disbursements of railways, have already been discussed by the writer in volumes referring specifically to such matters. The volumes in question embrace, incidentally, many of the regulations governing the business of railways, and as much of the information they contain might properly find lodgment in this book, it is not improper to explain here the reason of its omission.

Agents must attend to the comfort and convenience of travelers, and must give information when requested by them in a courteous and satisfactory manner.[1]

Agents will observe the deportment of trainmen toward passengers, and will report to the Superintendent any rudeness or incivility that may come under their observation.

### FREIGHT REGULATIONS.

The rules and regulations accompanying the freight tariffs of the various companies are more or less particular to recapitulate the circumstances under which freight will be received by them, and the extent of their responsibility for the property which they transport. Many of the regulations and exceptions are exceedingly pertinent, and in accordance with

---

[1]. "They must be courteous and respectful in their deportment to passengers, and if any agent is known to be otherwise, he will be reported to the Superintendent for misdemeanor, and, if the offense be repeated, be liable to suspension or dismissal. As much fault has been found with some of the sellers of the road for their want of courtesy, a strict observance of this rule is requested."—1854.

"He must take care that all the servants at his station behave respectfully and civilly to passengers of every class. He must take care that all the servants come on duty clean in their persons and clothes, and in the uniform supplied to them. Every exertion must be made for the expeditious dispatch of the station duties, and for insuring the safety of the public, and punctuality of the trains. The station master must report, without delay, to his superior officer, neglect of duty on the part of any of the company's servants under his charge, and forward to him particulars of any complaint made by the public."—*English Standard.*

good business usage and the laws governing common carriers. Many of them, however, possess no value whatever. *Glendower* could call spirits from the vasty deep, and so could *Percy*, but neither of them ever elicited any intelligent response. And so any body can frame rules and exceptions governing the carriage of goods and passengers, but only those in harmony with the responsibilities of common carriers possess any virtue further than that they may, perhaps, sometimes induce the patrons of a company to exercise greater care in particular cases, than they otherwise would. But this good is perhaps more than counter-balanced by their pernicious effect upon employés. In many cases the regulations that hold good in one section or state, have no binding force elsewhere. It is impossible, therefore, in a work of this description to classify or arrange them. Hence the writer has, as a rule, designed to omit all reference to them herein.

**DIRECTIONS TO AGENTS RECEIVING FREIGHT FOR SHIPMENT.**

They must not take a verbal order for the forwarding of freight, but must in each instance require shippers to furnish a shipping ticket. It must contain a description of the marks upon the freight, the consignments, name of route, also name of nearest railroad station (if destination is not located upon a railway line), etc. The ticket must be filed and preserved for future reference. .

The shipping ticket for articles, which the tariff directs must be transported at owner's risk, must read "owner's risk." The receipt given for the property must also read "at owner's risk." Agents must see that shippers understand the conditions on which such property is received by the company.

Agents will not receive freight unless it is marked with the address of the consignee in full. Initials are not sufficient.

They will not receive shipments of flour, wool, rags, hides, iron, and other articles which can not be fully marked with the place of destination and name of consignee, and which are, in consequence, liable to be mixed with other consignments of a similar description consigned to other parties, unless such shipments are branded, numbered, or marked, so that each package or consignment may be easily distinguished and accurately described in the way-bill. To insure correct delivery at destination, the brands, numbers, or marks on each package must be entered in detail on the way-bill; such freight as is liable to pilferage, especially from connecting lines, must be carefully handled, and agents must satisfy themselves that such property has not been re-coopered or pilfered, or damaged by wet at the time of its receipt by them; and, further, that it is in all respects in good order.

Freight should be forwarded as soon as possible after its receipt.

When buggies and carriages not boxed are shipped, agents must see that they do not contain any loose articles, such as cushions, harness, whips, robes, etc.; all such articles should be boxed and shipped separately.

They will decline to receive freight for re-shipment the charges upon which have been prepaid from point of shipment to destination, unless the money to prepay to destination is tendered with the property. This rule does not, of course, apply in those cases where freight is billed through.

Charges on perishable property must be prepaid or guaranteed by responsible parties.

Articles that are not considered worth the charge at forced sale will not be taken, unless such charges are prepaid.

When cooperage is required, packages are subject to a charge therefor.

### DIRECTIONS TO AGENTS — RECEIPTING FOR FREIGHT.

Whenever freight is received at a station for shipment, they must invariably issue a receipt therefor, correctly filled up and in conformity with the printed form of receipt provided. They must, moreover, in each instance say to shippers: "Your attention is directed to the conditions printed in this receipt, which are the conditions upon which your freight is received by the Company."

They must know from personal examination that they receive the property they receipt for, and that the marks upon such property correspond with the marks as described in the receipt; also that property is well packed and clearly marked.

They must be sure when receipting for freight that the receipt contains an accurate statement of all marks upon the packages, also that it states the destination of the property, gives the brand of flour, the marks upon bales of wool, cotton or rags, upon barrels of oil, hogsheads of tobacco, bars or bundles of iron, the mark or description of tag on each package of hides, etc., etc. As property is frequently packed for shipment in second-hand barrels or boxes, without the original marks being obliterated or erased, agents must be careful to see that all such marks are obliterated before receipting for the property.

In receipting for cars loaded by shippers, the receipt must read "shipper's count," "more or less," except when the right number of packages, measurement, weight, quantity, etc., etc., are known by the agent to be in the car.

When a receipt is given covering a variety of articles, such as a lot of household furniture, each separate piece must be properly marked, and a bill of particulars furnished by the consignor.

If a package is broken, the agent must ascertain if any loss or damage to the contents has accrued, noting the particulars upon the receipt and way-bill.

When freight that is liable to be damaged by the weather is shipped in open cars, it must not be received except at the owner's risk, and the receipt which is given must so state.

If no rate is inserted in the receipt issued by the agent, he must draw his pen through the blank space provided for inserting the rate.

All freight, except that loaded by shippers, must be checked before it is receipted for, the quantity or full number of packages being stated in each instance. The receipt or way-bill must not read "shipper's tally," or "more or less."

When freight is received in bad order from transportation companies or from any person whatever, agents must be careful to note on the receipt and the way-bill as well, the exact condition of the property. The term "bad order" or "b. o." must never be used. Packages received in bad order must be carefully weighed and the weight entered upon the receipt and way-bill.

They must not sign receipts agreeing to deliver property at any point beyond the terminus of the road, but may agree upon and insert the through rate when specially authorized.

When freight is contracted through to any point upon another line, agents must enter the through rate on the bill of lading or receipt, also each road's proportion of the through rate on the face of the way-bill, unless otherwise directed. When charges are advanced they must enter the amount advanced upon the receipt or bill of lading. Charges advanced on shipments of live stock must be entered upon the contract.

**RELEASES.**

Releases for household goods and for other freight of a similar character must be taken in duplicate; they must be signed by the shipper and witnessed by the agent or his assistant. The original release must be retained by the agent and preserved for future reference, but the duplicates must be attached to the way-bill and sent forward with the property. Agents must examine new furniture offered for shipment, and if they consider it is not packed in a manner to sustain the necessary handling while in transit, they must not receive it, unless a release is signed by the shipper in the same manner as for household goods.

DIRECTIONS TO AGENTS — LOADING AND UNLOADING FREIGHT.

Property belonging to different individuals must not be mixed in loading. Each lot must be kept separate. If goods are loaded in a car for more than one station, the goods to be unloaded first must be put into the car last. They must keep the freight for each station together, each lot of goods being kept by itself. They must see that goods in their charge are carefully handled, and loaded in such manner that no damage may occur in transit by leakage of liquids, chafing of bales, etc.; casks containing oils (other than coal), turpentine, tar, molasses, or liquors must be loaded on the bilge, and carefully blocked, bung up; they must be placed as far as possible from freight likely to sustain damage by any leakage that may occur in transit.

Freight must be checked as it is loaded and unloaded.

They must use great care in loading and handling coal oil; it must not be loaded with other freight that can be damaged by it. So far as practicable, it should be loaded in stock cars, the casks being placed on the head and well secured.

Freight for way points must not be loaded into cars containing through freight; freight must not be loaded into cars containing grain in bulk, nor must two kinds of grain be loaded in the same car, unless in sacks or barrels; nor must grain in sacks or barrels be loaded in cars with bulk grain.

When cars are chartered by shippers care should be taken to see that they are not overloaded.

They must not, under any circumstances, load merchandise, coffee, sugars, etc., into cars unfit for such property — notably in cars formerly used in transporting kerosene oil, lime, or other penetrating odors.

To save unnecessary hauling of cars and otherwise economize in their use, agents must never send a car with a small lot of freight when the same can be readily and quickly loaded after the arrival of the way freight train, provided there are cars in such trains into which the property in question may be loaded.

Kerosene, coal oil, naphtha, benzole or substances of a like combustible nature, must not be loaded nor unloaded through freight houses, except in the day time; nor must lights be allowed near such packages.

They must see that cars are loaded and unloaded promptly; that the rules for the collection of demurrage for the detention of cars are rigidly enforced; that chartered cars, or cars loaded with grain or other property, are not dangerously loaded, permitting none to leave their station in such condition, and finally that shippers are charged for the delay of cars held in consequence of being overloaded by them.

Agents are required to exercise especial care in securing the doors and windows of cars loaded with live stock.[1]

### CARE MUST BE EXERCISED IN LOADING FREIGHT.

" The proper loading of goods being a matter of so much importance, not only as regards the goods, but also as to the safety of the line, clerks in charge must give it their particular attention; for when it is remembered that, by the slightest neglect in loading and securely

---

[1] " Living quadrupeds are only forwarded from and to certain stations. The receiver or sender has to watch the unloading or loading and make the necessary arrangements for tying.

" Sick quadrupeds are excluded from forwarding, also such as may contribute to spread any contagious disease, according to the regulations of the board of health.

" A railroad company is not obliged to forward wild beasts.

" All shipments of other living quadrupeds have to be accompanied by some reliable persons, who must take their stand in the cattle cars. This is not necessary with smaller animals or fowls, if shipped in well ventilated cages or coops."
—*Regulations Austrian Roads,* 1877.

" On the arrival of horse-boxes or cattle-wagons at any station, they must be immediately cleaned out, so as to prevent damage to floors by wet straw, dung, etc., remaining on the wood; and every horse-box, wagon, and other vehicle must be thoroughly examined inside and out, so as to ascertain whether they are in a fit state to travel without liability of injury to the horses, cattle, etc. Should the horse-boxes be short of head-collars, the circumstance is to be reported immediately to the Superintendent."—*Gt. Nor. Ry., England.*

fastening the load of any one wagon in the goods trains, which are continually running on the line, a fearful accident may occur, it is impossible to overrate the necessity of the most pointed and constant attention being given by clerks in charge, loaders, and others, to satisfy themselves, before any train is permitted to start, that the load of every wagon is secured in a manner sufficient to sustain the oscillation of the train, and the necessary shunting to which it will be exposed.

"The clerk in charge, or some other person properly appointed by him, should carefully examine the loads of the wagons of the goods trains stopping at his station.

"After every care and vigilance has been exercised in loading, it will be impossible always to prevent the load being disturbed in a long transit; and it is, therefore, essentially incumbent upon the servants of all companies to examine with particular care all trains arriving from foreign lines immediately on their entering upon their respective railways. Should the load appear to be disturbed, the wagon must not be allowed to proceed until it has been carefully readjusted; and this is more especially necessary in the case of timber, cotton, wool, machinery, or other articles of a lengthy or bulky construction."[1]

1. Reg. Clearing House, Eng.

## DIRECTIONS TO AGENTS — DELIVERY OF FREIGHT.

After the delivery of goods to a company to be forwarded, they become the property of the consignee, and neither the name of the consignee nor the destination of the property must afterwards be changed, except under his instructions, or by due process of law.

When property is consigned and shipped to the care of a second party, the agent must deliver the same to the party in whose care it is shipped, unless the party to whose care it is consigned countermands the order in writing. When property is consigned and shipped to the "order" of a certain party, with instructions to "notify" a second party, agents must notify such second party of the arrival of property, but will only deliver on the written order of the party to whose "order" it is consigned, and on surrender of the bill of lading, which latter must be carefully filed for reference.

Care must be exercised to see that freight is properly delivered; except as provided in the above rule the consignee is the owner of the property so far as the common carrier is concerned, and is the only person to whom the carrier can safely deliver it.

When parties to whom freight is consigned are unknown or can not be found, the forwarding agent must be requested to advise consignors, and ascertain their wishes regarding its disposal.

### FREIGHT FROM AND TO STATIONS AT WHICH THERE ARE NO AGENTS.

" Freight consigned to stations where there are no agents; also to stations where there are ticket, but not freight agents, must be prepaid. The forwarding agent will way-bill the freight to the first station beyond its destination where there is an agent, but at the rates current to actual point of destination, noting, in ink, on the back of way-bill, underneath the filing, instructions to the conductor to put off the freight at its proper destination, and to deliver the way-bill to the agent of the station to which it is directed. This agent, at the end of each month, will make an abstract of such way-bills and forward the same, together with the original way-bills, to the freight auditor.

"At stations where there are no agents, or where there are ticket but not freight agents, conductors will receive freight, requiring from shippers memoranda containing full shipping directions, which they will hand to freight agent of first station beyond the point where the freight was received. Upon receipt of such memoranda, agent will make way-bill from his station, but at the rates current from actual point of shipment to destination on this line, noting on the face of the way-bill the point at which the freight was loaded. Agents will take such way-bills into their accounts same as if the freight was shipped from their station.[1]

### DIRECTIONS TO AGENTS — WAY-BILLING FREIGHT.

Freight must never be shipped without a way-bill, duly numbered and dated, and entered upon the station books.

The way-bill must be a correct copy, in every particular, as to consignment, route, destination, and number of articles, of the receipt held by shipper.

Agents must never bill freight as a "lot," but must enumerate each article.

When shipping perishable property, agents must note "perishable freight" in red ink on the outside of the way-bill.

If agents receive an order to add advanced charges after property has been delivered to the owner, and are unable to collect such charges, they will report immediately to the office giving the order, but will not alter the way-bill.

When property is loaded into cars of a passing train at way stations, agents must enter the initial and car number on the way-bill, and

---

[1]. Henry C. Wicker, 1878.

must be careful to make a like notation on the freight-forwarded book, immediately upon the departure of train.

### DIRECTIONS TO AGENTS IN REFERENCE TO SEALING CARS CONTAINING FREIGHT.

When necessary to open a car in a through train for the purpose of receiving or discharging freight, both seals must be cut by the agent, but the car must afterwards be resealed by him.

When opening a car, the seals on each side should be examined to see if they are alike; any discrepancies that may be discovered must be noted on the way-bill.

When it is necessary to open a car containing way-freight, the seals of such car must be cut by the agent opening it, but it must be resealed by the agent at the last station where freight is delivered from it preceding that where the transfer of conductors takes place.

They must remove the seals from both sides of cars when unloaded; at the end of each month the old seals must be transmitted to the company's storekeeper.

They must specify, in their daily reports, the number of each car received without a seal or having the seal broken, giving place of shipment, destination of contents, and any apparent derangement thereof; if the car is not for their station they must reseal it.

Box freight cars containing merchandise, must be locked and sealed when loaded, and agents must take a receipt for such cars from conductors.

They must examine the doors of loaded cars left at their stations, and see that they are sealed, whether the cars are intended for their station or not.

They must receipt to the conductor for cars left at their stations, noting on the receipt the numbers of those cars, if any, having imperfect or broken seals.[1]

FREIGHT AGENTS — MISCELLANEOUS RULES.

The classification of freight provides for the great bulk of the articles offered for transportation. Articles not enumerated must be charged in accordance with the class to which they are clearly analogous.

Very heavy articles, also articles light in weight but bulky in character, when not otherwise provided for, will be charged at such rates as the general freight agent may decide, when no agreement to the contrary is made.

It is expected that agents will give information as to different routes with which the road connects, when inquiries are made by patrons of the line, but will not endeavor to influence

---

1. The practice of sealing cars, as described in the foregoing rules, is not in general use upon railways.

shippers in favor of any particular route. It is their duty to maintain a strictly neutral position, unless otherwise expressly ordered.

Agents must not allow persons wishing information as to shipments from or consignments to their station, to have access to their books. Any information referring personally to an applicant should at all times be promptly and cheerfully given.[1]

All correspondence must be carefully preserved.

Letters and statements relative to the company's affairs must not be shown to shippers or others, or made known to any one, except so far as may be necessary for the guidance and instruction of the company's servants.

They must not advance charges upon property, unless such charges are incidental to its transportation.

They must take receipts for charges advanced, and must carefully file and preserve the same for reference when required.

Cars containing gunpowder, or freight of a like combustible character, should be conspicuously labeled with the name of the article with which they are loaded.

[1] " Persons not regularly in the service, or not about to travel by the trains, have not the right of access to the stations. The booking offices must be kept perfectly private, and the public and others must not have access behind the screen or counter, at any station Persons are not to be admitted to the station or offices, to learn the business, without the sanction of the General Manager."—*Gt. Nor. Ry., Eng.*

They must see that the doors and windows of loaded cars are kept locked ; the end doors of cars must be fastened on the inside. Grain doors must be carefully secured, in the place provided, except when they are required for use for grain in bulk.

When a car is left irregularly from a train at any station, prior to its reaching its destination, the agent at such station must advise the agent at the station to which the car is billed, as well as the Superintendent of the division, giving the number of the car, the number of the train leaving it, also the reason why it was left.

They must see that conductors certify to the correct delivery of property described on way-bills for freight delivered at points where there are no agents.

When cars containing merchandise or other property, except lumber, become disabled, the contents must be transferred, unless the car can be repaired so as to go forward within twelve hours ; cars containing lumber may be detained for repairs a reasonable time. Perishable property must go forward without delay.[1]

---

[1] "When cars are left at any way station in consequence of being out of repair, it shall be the duty of the agent where such car is left to send word immediately, either by telegraph or letter, to the Superintendent of car shop, or to the nearest local car repairer, stating what is necessary to repair it. If the car can not be repaired promptly, and it is found to contain perishable property, the agent will have the freight transferred immediately and sent forward to its destination."— 1863.

When a conductor fails to take all the cars that may be ready to go, he must give his reasons therefor to the agent. In the event such reasons are not considered satisfactory by the agent, he will forthwith report the facts to the Superintendent, giving the name of the conductor, the number of the engine and the number of cars in the train.[1]

A detailed report must be made, on the last day of each month, of all freight remaining uncalled for; it must describe the property, where from, name of consignee, condition of the freight, its value, and the amount of charges.[2]

### DIRECTIONS TO AGENTS IN REFERENCE TO FUEL.

They must not allow the stock of wood and coal to run short, and will promptly report any failure in the supply.

The wood intended for use by engines must be arranged upon the platform in such quantities (ranks) as may be required for use by engines.

They must keep the receptacles for coal

---

[1]. "Whenever he has loaded cars to send which any freight train declines to take, if in his opinion such train be not fully loaded, he will report the case to the Master of Transportation, giving the name of the conductor, engineman, and the number of cars in the train."— 1853.

[2]. "A monthly return of all unclaimed property in the goods or parcels department is to be sent to the Superintendent or Goods Manager at King's Cross."—*Gt. Nor. Ry., Eng.*

filled, ready to be dumped into the tenders of engines without delay.

They must require a ticket for the amount of wood or coal delivered to each engine; they must examine each ticket to see that it bears the number of the engine, and corresponds with the amount furnished. The tickets collected must be sent to the home office at the close of each month.

They must keep a record book of wood and coal consumed by engines; this book must be transmitted to the home office with the fuel tickets, at the close of the month; when examined and compared with the tickets it will be returned to the agent.

## DIRECTIONS TO AGENTS IN REFERENCE TO SWITCHES.

They will have charge of switchmen at stations, and will be held responsible for the position of switches; they must keep it in mind that a train may arrive at any moment, and must be prepared accordingly.[1]

They must see that switchmen properly signal all approaching trains.

The greatest care must be exercised in the

---

[1] "They (flagmen and switchmen) must be provided with a crowbar, shovel, sledge, spiking mauls, spikes, red and white lanterns, and with a flag-staff eight feet long, and have a white flag three feet square at one end and a red flag of the same size at the other end."—1854.

cleaning, trimming, and lighting of signal lamps, and agents will be held responsible for this work being efficiently performed.

When day and night switchmen are employed, they must not be allowed to leave their posts until relieved by each other, and the one going off duty must inform the one coming on of trains that are due but that have not arrived.[1]

Lamps of switches must be kept trimmed and in order, and must never be allowed to go out at night.[2]

Agents must see that switches are kept free from snow and other obstructions.

Switches must be set for the main track, and must be kept locked, except while being used.

### DIRECTIONS TO AGENTS IN REFERENCE TO TRAINS AND CARS.

All vehicles switched off at stations, as empties, must be carefully searched. The windows

---

[1] "When any one beat or post is covered for twenty-four hours by a day and night man, who relieve each other, the day will usually comprise thirteen hours, and the night eleven hours."—*Gt. Wes. Ry., Eng.*

[2] "He must satisfy himself that the signalmen at or attached to his station perform their duties in a proper manner by night as well as by day, and in order to maintain a proper supervision over the men in this respect, it will be necessary for him frequently to visit the signal boxes."—*Eng. Standard.*

of all empty passenger cars must be closed when they are standing on sidings at the stations.[1]

They are responsible for cars remaining at their stations; they must see that the brakes upon such cars are applied, and the wheels securely blocked so that they can not be moved by unauthorized persons, or blown by the wind, so as in any way to interfere with the safety of trains.[2]

Agents must see that tracks are kept clear and unobstructed, and they will not allow any train or engine to approach their station unless they can do so without danger. They must

---

1. " The windows of all empty compartments must be closed, not only while the carriages are standing at the stations, but also when the trains are running, immediately upon the compartment becoming vacant. The ventilators must be kept open." — *Eng. Standard.*

2. " The station-master must see that all fixed scotch-blocks at his station are kept across the rail ; that all safety-points are closed against the main line, when it is not necessary that they should be open for the purpose of shunting, and that all vehicles are placed within such scotch-blocks or safety-points. Facing-points not worked from a locking-frame must, in all cases, be securely fastened or held for the passage of trains.

" The station master, or person in charge, must take care that, while shunting wagons or other vehicles at stations or other places situate on inclines, in addition to screwing the van brakes tightly down, a sufficient number of wagon brakes are pinned down, and sprags or hand scotches used when necessary, to prevent the possibility of the train or any of the vehicles running down the incline. At such stations and other places a supply of sprags and hand scotches must be kept for the purpose. When wagons require to be shunted into incline sidings the trucks to be moved at one shunt must be limited to such a number as the engine can push up without going at a violent or excessive speed."—*Eng Standard.*

promptly report defective frogs or switches to the roadmaster.[1]

They are required to report accidents occurring to trains at or near their stations; all damaged cars or goods brought to or left at their stations, destined elsewhere, also, the amount of the damage, and how caused.

"When a horse is used on the railway, a man must, in all cases, have hold of its head, whether the horse is drawing vehicles or otherwise."[2]

### GENERAL DIRECTIONS TO AGENTS.

In the absence of a yard master the duties of that official are performed by the agent.

They have charge of the accounts, books, papers, buildings, sidings, grounds and property of the company, and of the property intrusted to it in the transaction of business at their respective stations, and will be held responsible for the safe keeping and proper care of the same, also for the efficiency of employés subordinate to them.[3]

---

1. "They will know personally, at least ten minutes before any regular train is due, and before leaving their stations at night, that the switches upon the main track are properly secured and locked, and that the cars upon their side-tracks, nearest the switches, have their brakes set, or their wheels well blocked." — 1863.

2. English road.

3. "Every station master or person in charge of a station is answerable for the security and protection of the office and buildings, and of the company's property there. He is also responsible for the faithful and efficient discharge of the duties

They must keep the buildings and grounds connected with their stations clean and in proper condition for the accommodation of passengers and the reception of freight, and must preserve order and system in and about their stations.[1]

They must keep their accounts and make their returns in such manner and form, and at such times as the accounting officer may direct.

They must keep the general rules and regulations of the company intended for the information of the public, governing the transportation of passengers and freight, posted in a conspicuous place in their depots.[2]

Agents are not allowed to be absent without leave from the Superintendent, except from illness, in which case they must immediately inform him of the fact. When absent, they

devolving upon all the company's servants, either permanently or temporarily employed at the station, or within its limits, and such servants are subject to his authority and directions in the working of the line. He is also responsible for the general working of the station being carried out in strict accordance with the company's regulations, and must, as far as practicable, give personal attention to the shunting of trains, and all other operations which in any way affect the safety of the line. He must always appear in uniform when on duty, if uniform be supplied to him."— *Eng. Standard.*

1. "When an engine or train of cars is left at the station over night, he will take general supervision and care of the same."— 1853.

2. "The notices connected with the company must not be stuck on the walls of the stations or offices, but are to be put on boards provided for that purpose; and all notices, last month's bills, etc., must be carefully removed when they cease to be needed." — *Gt. Nor. Ry., Eng.*

must leave their stations in charge of trustworthy and competent persons.

They must be careful that the company's stores are prudently and economically used, and that there is no waste of oil, fuel, or stationery, etc.[1]

They must use all proper means to secure traffic for the road, avoid giving offense, and act with a view of accommodating the public.

They must see that all orders of which they are cognizant are promptly executed.

They must promptly report to the Superintendent all deviations from the rules and regulations of the company, or anything that comes under their observation that is prejudicial to its interests, or that may interfere with the safe and economical working of the property.[2]

[1] 'The purchase of miscellaneous articles, or making of small bills, is strictly prohibited, except in cases of absolute necessity. Their necessary wants will be supplied by application to the Secretary of the Operating Department or Superintendent."— 1853.

[2] "They must report, without delay, neglect of duty on the part of any one at, or passing, their stations which may come under their observation."— 1854.

## CHAPTER X.

### GENERAL INSTRUCTIONS.

An employé can not become entirely familiar with the rules and regulations governing his duties except by acquiring knowledge of the duties of others.[1] This knowledge can not be acquired without an attentive perusal of the various rules and regulations; he will find something that interests him under all the various headings and sub-headings; it is impossible to accurately classify under different headings the duties of the various employés without endless reiteration. All the rules and regulations should therefore be studied.

One of the tests of an employé's fitness is the extent and accuracy of the information he possesses in reference to train and station service; this is especially the case with train and station officials. Each train official should be especially familiar with the duties of the various servants of the company connected with the train service, so that in the event of accident he may, if necessary, be prepared to

[1]. "All clerks in charge, inspectors, and foremen porters, are required to learn how to work the electric telegraph, and to keep themselves in constant practice, so as to be able to send messages in case of need."—*Gt. N. Ry., Eng.*

perform their functions. The same rule holds good in its application to employés at stations. No man is worthy of retention in the service, much less of promotion, who does not strive actively to acquire knowledge of his profession.

"All officers, clerks, and persons holding situations of trust will be required to find security for their faithful service, the amount and conditions of which security will be stated upon appointment."[1]

Employés must be sober, temperate men;[2] they must not accept gratuities, fees or perquisites;[3] they must devote themselves exclusively to the service of the company, attending diligently to their duties during the prescribed hours of the day or night, and they must reside wherever the interests of the company require.[4]

---

1. English Standard.

2. "Smoking while on duty is forbidden, and the use of intoxicating liquors as a beverage will be considered just cause of dismissal from the service of the company."—*A Western Road.*

"The proprietors of refreshment rooms are forbidden to supply spirits to any engineman, fireman, guard, or other servant of the company while on duty."—*Gt. Nor. Ry., Eng.*

"No instance of intoxication on duty will ever be overlooked."—1854.

3. "No person is allowed to receive any gratuity from the public, on pain of dismissal, and the compensation paid will cover all risks incurred, or liability to accident from any cause on the road."—1854.

4. "Each officer and man shall devote himself to the company's service, and he must serve when and wherever he is required, including Sunday if necessary, he being allowed for any extra work at his usual daily rate of compensation.

"If a guard or other servant should have two residences, he must make them both known at each station from whence he works."—*English Road.*

All property which they may find or which may come into their possession must be turned over to the authorized officer of the company to await the disposition of the owner.[1]

Employés must obey promptly instructions received from persons placed in authority over them, in conformity with the rules and regulations of the company.

Disobedience to orders, negligence, incompetency, or immorality renders a person unfit for retention in the service.[2]

Employés will not be permitted to absent themselves from their duty without the consent of the head of the department. Permission to be absent must be asked by employés through intermediate heads, when such employés are

---

[1] "All property which may be found on the line or premises of the company, by any man in their employ, shall be immediately handed to his superior officer, and by him to the agent at Blank street station, and entered by him in a book kept for that purpose. But should it be known that the property found had fallen from any particular train, it should be forwarded by the next train, or as soon thereafter as possible, to the station to which the train was proceeding, and notice thereof sent to the office at Blank street. Any man known to keep any property so found will be severely punished." 1854

"All property found by any servant of the company on any part of the premi-es must be immediately taken to the clerk in charge, in order that a proper entry may be made of the article in case of inquiry."—*Gt. Nor. Ry., Eng.*

[2] "Persons who disapprove of the regulations adopted, or are not disposed to aid in their enforcement, are requested not to remain in the employment of the company."—1854.

"And they will inquire into and punish instances of immoral or loose conduct on the part of any of their servants."—*English Road.*

not directly responsible to the chief officer of the department, or the next official in rank.[1]

All orders and instructions must be carefully preserved and filed for future reference.

Employés are required to exercise a wise discretion and economy in the use of the company's material intrusted to their care.

Any neglect of the storekeepers to furnish employés with materials, blanks, books, and other supplies, in such quantity and of such quality as may be required to do the business of the company in an expeditious and economical manner, must forthwith be reported to the Superintendent, or the department officer interested.

Articles required for use by employés such as lamps, keys, flags, axes, saws and other classes of material, will not be allowed without

---

[1] "Men absenting themselves without leave, and prevailing on others to supply their places, will subject themselves and all parties concerned to a heavy fine. Any man absenting himself without having a proper 'leave of absence ticket,' will be fined $1.25, as though he were absent without leave.

"In case of extra business, of sickness, or unavoidable cause of absence of any servants (excepting clerks) the clerk in charge is immediately to provide for the proper performance of the duty by appointing some temporary substitutes, but he is responsible for selecting men of good character, sober, honest, and intelligent, and capable of undertaking the office. With a view to such temporary appointments, it is desirable that the character and eligibility of some proper persons from time to time be previously ascertained."—*Gt. Nor. Ry., Eng.*

the return of the corresponding article previously in use.[1]

Employés intrusted with keys to switches or cars are required to receipt for them, and must not let them go out of their possession.

Persons leaving the company's service must deliver up any property belonging to it intrusted to their care. If the property shall have been improperly used or damaged, a sufficient amount must be withheld from the pay of the person to make good the loss suffered.[2]

Employés will be held responsible for injury occasioned to persons or property by their negligence or misconduct, also for all moneys that may come into their possession, and the company reserves the right to reimburse itself for any expense it may be put to in consequence of any negligence, misconduct or improper action upon the part of an employé, by withholding the pay of the person or persons in fault.[3]

---

1. "Broken lamps must be sent to the lamp room, King's Cross, for repairs, accompanied by the proper way-bill, a duplicate at the same time being sent to the Superintendent of the line."—*Gt. Nor. Ry., Eng.*

2. "And if he occupies one of the company's houses, he shall immediately remove his furniture from it, and put the house into as good condtion as when he received possession of it."—1854.

3. "In the event of any misconduct or suspicion of irregularity of the servants, it is competent to the district agents or clerks in charge to suspend them, reporting the circumstances immediately. The pay of all clerks, guards, policemen, porters, and others, will be stopped from the moment of their being suspended ; and the pay will not be allowed except in

Persons in the employ of the company are forbidden, while upon duty, from entering into altercation with other persons, no matter what provocation may have been given.

Employés in places of trust must report any misconduct or negligence affecting the interests or safety of the property which may come to their knowledge.

Employés are not allowed to use the credit of the company without the written authority of the Treasurer of the company.

The pay of employés absent from duty will be stopped, unless otherwise directed by the head of the department.[1]

the event of entire acquittal of the charge for which the man was suspended. The company reserve the right to deduct from pay any fine imposed for neglect of duty, or otherwise, which (in the event of pecuniary loss to the company not being entailed thereby) will be appropriated to a benevolent fund."
—*Gt. Nor. Ry. Eng.*

1. "A clerk, in case of continued absence on account of illness, is not entitled to pay for more than a fortnight during such absence, except under the special sanction of the board, to whom application must be made through the Superintendent of the line, who will decide whether the case be one he can properly recommend for consideration; but as a sick fund is now established to which all persons in the service are eligible, and which, for a small weekly payment, provides medical attendance for the contributors, their wives and families, a weekly allowance in sickness, and funeral allowance in case of death, clerks are recommended to subscribe to it, and thus render themselves, in a much greater degree, independent in case of sickness or other unavoidable calamity befalling themselves, or their wives or families.

"Every guard, policeman, and porter, is required to become a member of the sick fund established by the company, and to pay his subscriptions regularly out of the wages he receives by deduction from the pay-bill, or otherwise."—*Gt. Nor. Ry., Eng.*

When instructions are not understood, or when the course to be pursued admits of doubt, employés must so act as not to compromise the safety of the property or endanger the lives of passengers or others, seeking of the proper officer, on the first opportunity, the explanations they require.

Employés connected with the train or station service must have in their possession a copy of the schedule and the rules and regulations forming a part of it.

## CHAPTER XI.

### REGULATIONS OF THE AUSTRIAN RAILWAYS GOVERNING THE PASSENGER SERVICE.[1]

Railroad employés must treat the public in a polite, modest, and business-like manner, and must be obliging, as far as the service will allow. They must render all the services required of them gratuitously; it is prohibited them to accept any compensation from the public; employés are not allowed to smoke when they are on duty.

The public must conform to the wishes of employés, who are to be recognized by a uniform.

Differences between the public and employés are to be decided by the station-manager, or, on the road, by the conductor.

Complaints must be made to the officers, either verbally or in writing, or must be entered in a book which can be found for this purpose at each station. The managers must

---

[1]. Laws regulating the management of railroads in the Kingdom and provinces represented in the Council of the Empire and by-laws given the 25th of July, 1877. Translated by M. Blanque.

The regulations of the German roads are, in many respects, the same as those of Austria.

give an answer, at an early date, to all complaints, to which must be added the names and residence of complainants. Complaints in reference to an employé must specify the name, number, or uniform of the latter.

The public are to have admittance only to such parts of the depot and railway grounds as are always kept open, or are open temporarily for the convenience of the public. Walking on the tracks or roadway is not allowed, except to those who possess the right in accordance with the regulations of the railway police.

Forwarding of passengers, quadrupeds, etc., can be refused if uncontrollable, or circumstances should arise, or superior power interfere, or if the regular means for forwarding should be insufficient.

Payments must be made in current gold and silver coin, excepting fractional currency, in accordance with the rates published by the railroad management.

The forwarding of passengers is regulated by the time-tables hanging on the wall at all stations. The time-table also states what classes of cars the respective trains haul. The running of special trains is left to the consideration of the management. The station clock regulates the time for starting trains.

The prices of tickets are given in a tariff posted up in a conspicuous place at each station.

Tickets secure seats in the respective classes as far as there are such seats. If a passenger can not obtain such a seat as the ticket issued to him entitles him to occupy, and if there is no vacant room in a higher class, he is at liberty to exchange his ticket for one in a car of a lower class, the difference in price being refunded to him, or he has the right to ask for the return of his money, thus renouncing the obligation of being forwarded. Those passengers who are in possession of through tickets must be disposed of first.

Each ticket sold must show the names of the stations between which it is good, also the price of the class which the passenger intends to travel in;[1] finally, the time or the train for which the ticket is good.

The time or train for which a ticket has been issued must be stamped upon it, so that the purchaser can see at a glance whether it answers the purpose or not. The passenger has the right to stop at an intermediate station and take another train of corresponding grade on

[1] "Private servants (male and female) accompanying gentlemen's carriages by ordinary trains, are allowed to travel in or upon such carriages with second-class tickets; if by the third-class train, with third-class tickets; but this privilege does not extend to any other than servants. Servants when accompanying their masters traveling by express trains, are charged second-class express fares ; but this can only be the case if such servants are properly identified by their masters or mistresses who may be traveling with them."—*Great Northern Railway of England.*

the same or following day; but in such case, after alighting from the train, he must present the ticket to the station-manager to have its validity extended. The time granted on trip, or return tickets, can not be extended.[1]

Prices are reduced and tickets issued for children under ten years, and should there be any doubt about their age, the decision of the revising officer is final. No fare will be paid for small children carried in arms, or who occupy no extra room.

The exchanging of tickets of a lower for a higher class will not be allowed within ten minutes of the starting time of trains, and will not be allowed in any event unless there are unoccupied seats in the class desired. When tickets are exchanged the difference in price must be paid. At intermediate stations such exchange will not be allowed except when an additional ticket is purchased to the place of destination, the value of which added to the

---

[1]. "A return ticket is granted solely for the purpose of enabling the person for whom the same is issued to travel therewith to and from the stations marked thereon, and is not transferable. Any person who sells, or attempts to sell, or parts, or attempts to part, with the possession of the return half of any return ticket in order to enable any other person to travel therewith, is hereby subjected to a penalty not exceeding forty shillings, and any person purchasing such half of a return ticket, or traveling or attempting to travel therewith, shall be liable to pay the fare which he would have been liable to pay for the single journey, and shall, in addition thereto, be subjected to a penalty not exceeding forty shillings."—*Eng. Standard.*

value of the ticket first purchased equals the price of the higher classed seat desired.

Particular seats can not be sold or reserved in advance. Employés have the right, and on demand of passengers are obliged to point out seats to the latter. Ladies traveling alone must be seated in separate ladies' coupé when they desire it.

A separate ladies' coupé must be provided in all trains for passengers of the second and third class. This distinction will be modified as necessity requires when cars are constructed after the American system.

At all stations the waiting-rooms must be opened at least one hour before the train leaves.

On entering the waiting-room the passenger, if desired, must exhibit his ticket, also, on entering the car.

During the journey passengers must retain their tickets until the same are collected.

Any passenger who shall not be in possession of a valid ticket must pay a fine double the amount of the fare for the distance traveled, and any passenger who, when going on board of a train, tells the conductor thereof that he (the passenger) was too late to buy a ticket, and is allowed to stay on board of such train, must pay, in addition to the fare, 50 kreutzer.[1]

---

[1] "The guard must not allow any passenger or parcel to be conveyed by the train unless properly booked; and if he has reason to suppose that any passenger is without a ticket,

If the passenger refuses to pay such fine he can be put off the train.

The sign to enter the cars is given by two strokes of the bell.

No one is allowed to get on board the train after the sign to start has been given by the whistle of the locomotive, and any effort to do so is punishable.

A passenger who misses the train in the manner described has no claim for the refunding of his fare or for indemnification of any kind. But he has the right to use the ticket in his possession on the next day upon a train of the same class, but the ticket must be extended by the station-keeper. This extension can not be applied on return or round-trip tickets.

On arrival at a station, the name of same and length of sojourn, and any changing of cars must be called. After the train has stopped, the doors of the cars which have this station as the point of destination will be opened. The doors of other cars will only be opened if desired.

Any one leaving his seat without first securing its retention must take another one in the event it is occupied during his absence.

If a train is stopped outside of a station on

or is not in the proper carriage, he must request the passenger to show his ticket, reporting to the station-master or person in charge, any irregularity he may detect. When a passenger is desirous of changing from an inferior to a superior class of carriage, the guard must have this arranged by the station-master or person in charge."—*Eng. Standard.*

account of some obstacle, no one will be allowed to leave the cars without the conductor's consent. Passengers must not stand upon the track, and must resume their seats upon the first signal of the whistle. The signal to start is three blasts of the whistle; any one not on board when the signal to start is given will be excluded.

While the train is moving, no one is allowed to look out of the cars, lean against the doors, or step on the seats.

If objection is made by one passenger only, the windows on the windward side can be closed.

Only employés have the right to open the doors for entering and leaving the cars; no stepping off the cars is permitted until the train has come to a full stop.

Every passenger must keep at a distance from the rails and machines and must leave the depot in the direction prescribed.

Any damage done to the cars, by passengers, must be paid for according to the indemnification tariff, and employés are empowered to make collections at the time in accordance with such tariff.[1]

[1] "Any person who willfully cuts or tears any lining or window strap, or curtain, removes or defaces any number plates, or breaks or scratches any window of a carriage used on the railway, or who otherwise, except by unavoidable accident, damages, defaces, or injures any such carriage, or any station, or other property of the company, is hereby subjected

Claims can not be made on account of delayed trains.

The abandonment or interruption of a train during a voyage, only justifies a claim for the amount of the fare for the distance not traveled by the passenger.

If connection with another train should have been missed and superior power has not been the cause, the passenger, if he takes the next return train to his starting point, is entitled to have the amount of both fares refunded to him on proof of his claim. Such passenger, however, to secure his claim, is obliged, on arrival of the belated train, to report to the station-keeper and present his ticket. The latter must confirm the delay and the station-keeper of the starting point has also to certify to the time of the passenger's return. In case interruption to a voyage is occasioned by the elements, or obstacles have damaged the railway, arrangements must be made to forward passengers in the best manner possible. Irregularities must be made known to the public by visible placards posted at the different stations.

Dogs and other quadrupeds are not allowed in the cars; lap-dogs are excepted from this rule in those cases where no objections are made by passengers.

to a penalty not exceeding five pounds, in addition to the amount of any damage for which he may be liable."—*G. W. Ry., Eng.*

Smoking is allowed in all classes of cars, but in the event there is no smoking coupé of the first class in the train, smoking will not be allowed in coupés of the first class when passengers object. Every passenger train must contain second-class, and, if possible, third-class coupés in which smoking is prohibited. Tobacco pipes must be sufficiently covered.

Baggage containing combustible articles, liquids or other articles which might do injury, especially charged guns, gunpowder, easily inflammable preparations and things of such nature, are not allowed in the passenger cars. Employés are empowered to examine such articles closely. Any one disregarding this rule is responsible for any injury caused, and is also subject to a fine according to the regulations of the railway police. Huntsmen must have a special permit.

Transgressions of the rules prescribed, acting in opposition to employés' wishes, indecent behavior or drunkenness, will lead to the exclusion of the person or persons in fault from the cars, and in such cases fare will not be refunded.

Drunken persons will not be allowed admittance to the waiting rooms or cars, and must be ejected when they gain access thereto.[1] If

---

1. "Any person found in a carriage, or elsewhere upon the company's premises, in a state of intoxication, or using obscene

ejected during the voyage, as provided by this rule, or after having surrendered baggage to the company for forwarding, the person or persons ejected are only entitled to have their baggage delivered at the station to which it was originally directed.

or abusive language, or writing obscene or offensive words on any part of the company's stations or carriages, or committing any nuisance, or otherwise willfully interfering with the comfort of other passengers is hereby subjected to a penalty not exceeding forty shillings, and shall immediately, or, if a passenger, at the first opportunity, be removed from the company's premises."—*G. W. Ry., Eng.*

## CHAPTER XII.

### A CHAPTER DEVOTED TO THE RULES AND REGULATIONS OF THE GREAT ENGLISH ROADS.[1]

*General Regulations.*

Every person employed by the company must devote himself exclusively to their service, residing at whatever place may be appointed, attending at such hours as may be required, paying prompt obedience to all persons placed in authority over him, and conforming to all the rules and regulations of the company.

Although the rules and regulations given under different heads are made specially for the observance of the servants employed in doing the work required by such rules and regulations, yet every such person must make himself thoroughly acquainted with them, and will be held responsible for a knowledge of, and compliance with, the whole of them.

Every servant is required to assist in carrying out the rules and regulations, and must immediately report to his superior officer any infringement thereof, or any occurrence affecting the safe and proper working of the traffic which may come under his notice.

[1]. Clearing-house Standard, 1877.

The address of each person employed in the working of the railway must be registered at the station to which he is attached, or at which he is paid, and must be posted in the station-master's office, so that, if required in cases of emergency, the men may be readily found. Any change of address must be notified to the station-master, in order that the record may be kept perfect.

No servant is allowed, under any circumstances, to absent himself from duty, or alter his appointed hours of attendance, or exchange duty with any other servant, without the special permission of his superior officer. In case of illness, he must immediately report the circumstance to his superior officer.

Every person receiving uniform is to appear in it, when on duty, clean and neat, with the number and badge perfect, and if any article provided by the company shall be damaged by improper use, he will be required to make it good. No servant is allowed to convert to his own use any article, the property of the company, and, if guilty of such misconduct, he will be severely punished. The conduct of all servants must be prompt, civil and obliging. They must at all times afford every proper facility for the business to be performed, be careful to give correct information, and, when asked, give their names without hesitation.

All officers, clerks, and persons holding situations of trust, will be required to find security for their faithful services, the amount and conditions of which security will be stated upon appointment.

No officer or servant of the company is allowed to travel on the railway, unless provided with a proper ticket, or free pass; nor is he allowed to ride on the engine, or in the brake van, or in any vehicle in which luggage or parcels are conveyed, unless in the execution of his duty, without written permission from the properly authorized officer of the company.

No guard, engine-driver, fireman, signalman, policeman, porter, or other servant of the company, while on duty, is allowed to enter a station refreshment-room, except by permission of the station-master, or person in charge of the station.

No money or gratuity in the shape of fee, reward, or remuneration, is allowed to be taken from passengers, or other persons, by any servant of the company, under any pretense whatever, even although the regular hours of duty shall have expired.

No servant of the company is allowed to trade, either directly or indirectly, for himself or others. The company reserve the right to punish any servant, by immediate dismissal, fine, or suspension from duty, for intoxication, disobedience of orders, negligence, misconduct, or absence from duty without leave, and to deduct from the pay of their servants and retain the sums which may be imposed as fines, and also their wages during the time of their suspension, or absence from duty from any cause.

No servant is allowed to quit the company's service without giving the month's notice required by the terms of his engagement.

When a man leaves the service, he must immediately deliver up his uniform and all other articles belonging to the company, and no money due for wages to any man leaving the service will be paid until his clothing, book of rules, lamps, flags, tools, detonators, and all other articles, the property of the company, which may have been supplied to him, shall have been delivered up in accordance with the company's regulations. If not delivered up, or if any article be missing, or be damaged by improper use, the cost of such articles, or of the repair of such damage, shall be a debt due from the man to the company, and may be deducted from any pay then due, or, if such pay be found insufficient to meet the claim, will become a debt recoverable at law.

All testimonials and letters of recommendation will, if required, be returned by the company at the time the person whom they concern leaves the service; except such as are addressed to the company or their officers.

All servants must exercise proper care in getting between vehicles for the purpose of coupling or uncoupling them.

No trespassing upon the railway must be allowed, and no person must be permitted to walk on the line, unless provided with written or printed permission to do so, signed by a properly authorized officer of the company. In the event of any person trespassing, and refusing to quit when requested to do so, the name and address of such person must be obtained, and the circumstances reported to the nearest station-master.

Special trains or engines have frequently to be run without previous notice of any kind, it is therefore necessary for the staff along the line to be at all times prepared for extra trains or engines.

The safety of the public must, under all circumstances, be the chief care of the servants of the company.

Wherever the term "Main Line" is used, it means the running line of any railway, or branch. Whenever the word "Train" is used, it must be understood to include "Light Engine," i.e., engine without a train.

Wherever the words "Goods Train" are used, they must be understood to include "Goods, Mineral, Cattle, and Ballast Trains."

CONDITIONS UNDER WHICH PERSONS ARE ADMITTED TO THE SERVICE — SECURITY — PRIVILEGES — COMPENSATION — ETC.[1]

A candidate as an experienced clerk must possess railway experience, or experience in other traffic equivalent thereto.

The salary, not exceeding $400[2] per annum, is fixed on appointment.

A candidate as a junior clerk must have attained eighteen and must not exceed twenty-three years of age.

The salary on appointment and

| | | | | | | | |
|---|---|---|---|---|---|---|---|
| For the | 1st year, is | - | - | - | $5 25 | per week. |
| " | 2d | " | - | - | - | 5 50 | " |
| " | 3d | " | - | - | - | 5 75 | " |
| " | 4th | " | - | - | - | 6 00 | " |
| " | 5th | " and until promoted, | | | | 6 25 | " |

If employed in London, but during such employment only, $1.00 a week is allowed in addition to the salary.

A junior clerk is eligible for promotion only on a vacancy occurring, and upon the head of the department in which he has been employed, and the General Manager, recommending him as qualified to fill the same.

A candidate as a lad clerk must have at-

---

1. Gt. Nor. Ry., England.

2. I have taken the liberty here, as I have elsewhere herein, when I thought proper, of reducing the foreign currency to the American standard.—*M. M. K.*

tained fifteen and must not exceed eighteen years of age.

The salary on appointment and

| For the | 1st year is | - | - | - | $2 50 per week. |
|---|---|---|---|---|---|
| " | 2d | " | - | - | - | 2 75 | " |
| " | 3d | " | - | - | - | 3 25 | " |
| " | 4th | " and until promoted, | 4 00 | " |

A lad clerk is ineligible for promotion to be a junior clerk until he is eighteen years of age, and then only upon a vacancy occurring, and upon the head of the department in which he has been employed, and the General Manager, recommending him as qualified to fill the same.

All clerks, without reference to their standing in the service, are allowed $1.00 a week in addition to their pay, when employed wholly on night duty.

Written application at the end of each year of service must be made to the directors through the medium of the Superintendent of the line, or chief of the department in which the clerk is engaged, for the authorized increase of salary, and failing such application at the proper time, increased pay will be allowed only from the date at which it is eventually made. This rule applies also to the police and porters.

A candidate as a clerk will undergo a strict examination as to his qualifications, in proportion to his age; he will be required to show a good handwriting, suited for accounts and correspondence, and that he has a competent knowledge of mercantile arithmetic; and he must be in a good state of health.

The candidate must, on attending at the Secretary's office to be examined, produce testimonials of character.

In the case of an experienced clerk, and of a junior clerk who has been before employed, first, from his last employer; second, one from each of two housekeepers of undoubted respectability.

In the case of a lad clerk, and of a junior clerk who has not been before employed, first, from the head master of the school in which he has been educated; second, one from each of two housekeepers of undoubted respectability.

The nomination, with the particulars of the examination and the testimonials, will be submitted to the directors on the candidate appearing before them, and who will decide whether he be qualified and a proper person to be appointed.

The name of a clerk, on appointment, will be added to a list, from which he will be summoned in turn for duty as a vacancy occurs, provided he has in the meantime given security; but should he, on being summoned, refuse or neglect to join, his name will be struck out of the list, and he can not afterwards be re-admitted to the service.

A clerk must, immediately on appointment, give security to the amount of two years' salary, or in not less than $500, through the medium of one of the undermentioned guarantee societies, and he can not subsequently, under any pretense whatever, be allowed to change from the society first selected.

(Here follows the list of guarantee companies.)

The railway company pays the premium in the case of a clerk whose salary does not exceed $5.25 per week or $6.25 per week without allowances.

A candidate as a porter must be five feet seven inches in height, without his shoes. He must not be less than twenty-one, and must not exceed thirty-five years of age. He must be able to read and write, and be generally intelligent; free from any bodily complaint, and of a strong constitution, according to the judgment of the surgeon by whom he will be examined, who will report whether he is "fit" or "unfit." The police are selected from this class.

The candidate must produce testimonials of character from his last employer, and one from each of two housekeepers of undoubted respectability, and if he has been in any public service also a certificate of good conduct during such employment; these, with the nomination, will be submitted to the directors on the candidate appearing before them, and who will decide whether he be a proper person to be appointed.

The pay of a porter is, on entering, and

|  | In London. | In Country. |
|---|---|---|
| For 1st year, | $4 25 per week. | $4 00 per week, |
| " 2d " | 4 50 " | 4 25 " |
| " 3d " and until promoted, | 4 75 " | 4 50 " |

provided a fine be not incurred in the interim, in which case increased pay is allowed only after twelve months' service from the date of such fine.

A candidate as a lad porter must not be less than fourteen, nor exceed seventeen years of age. He must be able to read and write, and be generally intelligent, free from any bodily complaint, and of strong constitution, according to the judgment of the surgeon by whom he will be examined, who will report whether he is "fit" or "unfit."

The candidate must produce testimonials of character from the school at which he has been educated, and one from each of two housekeepers of undoubted respectability. These, with the nomination, will be submitted to the directors on the candidate appearing before them, and who will decide whether he be a proper person to be appointed.

The pay of a lad porter is, on entering, and

For the 1st year - - - - $1 75 per week.
" 2d " - - - - 2 00 "
" 3d " - - - - 2 25 "
" 4th " - - - 2 50 "
" 5th " - - - - 2 75 "
" 6th " - - - 3 00 "
" 7th " and until promoted, 3 50 "

A lad porter on attaining twenty-one years of age, and not before, is eligible for promotion to be a porter, but he can then become a porter only after being passed by the surgeon and the directors, as in the case of a new appointment, want of height (under five feet seven inches) not being, however, a disqualification.

All appointments are made on the distinct understanding that the parties hold themselves in readiness to proceed to duty immediately on being summoned, their pay being allowed from the date of employment, that they reside wherever required, and that they will join and become members, on being so required, of any provident or benevolent society established or to be established in connection with the company, and abide by all the rules and regulations * * * or otherwise given them for their guidance.

The rules of the Sick and Funeral Allowance Fund are furnished to every porter on appointment.

Station-inspectors, $6.25 and $7.50 per week,

*Trains and Stations.* 249

according to the class of station, with house, or an allowance of $1.25 per week in lieu.

| | | | | |
|---|---|---|---|---|
| Pass. Guards (con.) | 1st class, | Chief Guard | $7 50 per week. |
| " " | 1st " | Under Guard | 6 87 | " |
| " " | 2d " | Chief Guard | 6 75 | " |
| " " | 2d " | Under Guard | 6 25 | " |
| Goods and cattle Guards, | | Chief Guard | 7 50 | " |
| " " | | Under Guard | 6 87 | " |
| Mineral Guards, | - | - | - | - | 5 75 | " |

All guards when required to sleep away from home, receive twenty-five cents per night additional.

| | | |
|---|---|---|
| Police—Ordinary, - - - - | $4 25 per week. |
| " Signalmen at Junctions and Pointsmen in London, | 5 00 | " |
| " In the Country, - - - | 4 75 | " |
| " Gatemen at level street crossings, | 4 75 | " |
| " Gatemen at level r'd station crossings | 4 25 | " |

Gatemen provided with a house by the company, are to have coals free, and to pay sixty-two cents a week rent, but if they open the gates by night in addition to the day work they are to have the house rent free, as an equivalent for the night work.

| | | |
|---|---|---|
| Porters in London, - - - | $4 25 per week. |
| " in the Country, - - - | 4 00 | " |
| Foremen Porters in London, - - | 5 25 | " |
| " " the Country, - - | 5 00 | " |
| Mineral Foremen Porters in the Country, | 5 25 | " |
| Shunters in London, - - - - | 4 75 | " |
| " the Country, - - - | 4 50 | " |
| Luggage Stowers and Loaders, - - | 4 75 | " |

Police and porters are to receive an advance of twenty-five cents per week each year for two years, beginning on the day when they shall have completed a year's service, if not punished in the interval.

Foremen porters, signalmen or pointsmen, gatemen at level street crossings, shunters and loaders are to be advanced under the same rule, twenty-five cents per week each year for two years, from which their only increase will be by promotion to a superior foremanship at $6.25, which is a fixed rate of wages, or to the situation of guard or inspector.

In case of promotion, men who have been advanced under above rule are to carry with them and continue to enjoy their advance, unless the promotion is to a grade paid at a fixed rate of wages, when it will cease.

Signalmen, at the expiration of every half-year of good service, without punishment, will receive a premium of $12.50.

As soon as any fine or punishment for misconduct shall be registered against any servant of the company, the previous period of the current year's service for increase of pay or premium becomes forfeited, and the year can only be reckoned from the date on which he was punished.

## THE UNIFORMS REQUIRED AND THE REGULATIONS INCIDENT THERETO.[1]

All servants of the company to whom uniform is allowed are required to wear it while on duty. The uniform of servants clothed by the company is as follows, for twelve months:

For station-inspectors and guards, a great coat, a frock coat, waistcoat, two pairs of trousers, two red neckerchiefs, and hat or cap; for policemen, a great coat, a dress coat, two pairs of trousers, cape and hat; for porters, a jacket, waistcoat, two pairs of trousers, two red neckerchiefs, and cap.

Foremen porters and shunters have a cape in addition. Authorized laborers receive two blue "slops," and red neckerchiefs.

Uniforms will be issued as follows: To the inspectors and guards, a top coat once a year, and a frock coat once a year. When a second of either garment is issued the first may be retained, but when a third is served out the first issued is to be given back; when the fourth is issued the second to be given back, and thus two of each garment will remain in their possession. The trousers and hats or caps remain in the possession of the men, except that, when they leave the service, two

[1]. Gt. Nor. Ry., England.

pairs of trousers must be given up, with all other clothing and appointments.

To the police, a great coat and cape every two years; on receipt of new ones the old ones must be given up. The dress coats in use when the second coats are supplied are allowed to remain in possession of the policemen until a third is issued; they are then required to give up No. 1, keeping Nos. 2 and 3; when No. 4 is issued No. 2 is to be given up, and so on, two dress coats remaining in the possession of the men. Hats and trousers remain in possession of the men, except that when they leave the service, they are required to give up two pairs of trousers, with all the other clothing and appointments.

Porters are subject to the police regulations as to their jackets and waistcoats. When the second jackets and waistcoats are issued, the first are retained by the men; when the third are issued the first are given back, and so on. The trousers remain in possession of the men, except that, when leaving the service, they are required to give up two pairs of trousers, and all the other appointments of clothing. The capes are issued once in two years, the caps and neckkerchiefs yearly; on receipt of a second cap or cape the first is to be given up.

GENERAL REGULATIONS FOR WORKING THE ABSOLUTE BLOCK SYSTEM ON A DOUBLE TRACK ROAD.[1]

The signaling of trains on the block telegraph system does not in any way dispense with the use of home, distant, starting, hand, or fog signals, whenever and wherever such signals may be requisite to protect obstructions on the railway. The object of the system of electric train signaling is to prevent more than one train or engine being between any two signal stations on the same line at the same time. This is accomplished by not allowing any train or engine to leave a signal station till the previous train or engine has been signaled as having arrived at or left the signal station next in advance.

The block signal instruments and bells are exclusively for the signaling of trains, and must not, under any circumstances, be used for conversing, nor for any other purpose than block-working, in strict accordance with the company's regulations, and they must only be used by the signalman, or other person specially appointed for the duty.

The signal boxes at which the block telegraph working is in operation, are furnished with instruments to signal for each line of rails,

---

[1]. English Clearing House Standard.

and the system under which these instruments are to be worked, and the mode of indicating the description of approaching trains, will be laid down in the code of regulations supplied to signalmen or exhibited in the signal boxes for the guidance of the persons in charge.

On those portions of the line worked on the absolute block system, a second train or engine must not be allowed to enter a section until the preceding train or engine has been signaled as having passed out of the section, except under the circumstances specified in rules " A " and " B," further on, to meet cases of train or telegraph failure. The danger signal must be exhibited at both the home and distant signals[1] to protect trains or engines standing at stations or intermediate signal boxes, and when any train or engine has gone forward into the onward section, the starting and advanced starting signals (where such are provided), which control the entrance of trains and engines into such sections, must also be put to, and kept at, " danger," until telegraphic information has been received from the signal box in advance that the preceding train or engine has passed out of the section. So long as the starting signals stand

---

1. The "home" signal or semaphore is located in the immediate vicinity of the station ; the "distant" signal is, however, located further away. It is usually worked (by means of a chain running along the ground) by the person who operates the " home " signal.—*M. M. K.*

at "danger," the home and distant signals must also be kept at "danger," except on the near approach of a train which has to stop at the station, when, after the speed of the train has been reduced so as to admit of its stoppage at the platform, the home signal may be taken off to admit the train, but the starting signal must be kept at "danger" until the line is clear to the next signal station ahead.

Unless special instructions are given to the contrary the line must be considered clear, and the signal "line clear" be given immediately the last vehicle (with tail-lamp attached) has passed the home signal post, except during foggy weather or snow-storms, when the signal "line clear" must not be sent to the station in the rear until the train or engine that has stopped at the station has passed the home signal, and is proceeding on its journey, or has been shunted into a siding clear of the main line.

Should it become necessary to block a section, in consequence of a breakdown obstructing the line, or other circumstances taking place rendering it imperative that any approaching train should be stopped, the signalman at the station where the obstruction takes place must use the means authorized by his regulations for preventing any train leaving the post in the rear.

Should there be reason to suppose that both

lines are fouled, the signalman must, without any delay, block the lines in both directions.

No obstruction must be allowed outside the home signal until the signalman on duty has carried out the prescribed regulations to prevent any train leaving the signal station in the rear.

If a signalman observe anything unusual in a train during its passage, such as signals of alarm by a passenger, tail-lamp missing or out, goods falling off, a vehicle on fire, a hot axle-box, or other mishap, he must give the station in advance the signal to "stop and examine train," and the signalman at the station in advance must acknowledge such signal, and instantly put on the danger signals to stop the approaching train. Where practicable, the signalman must also telegraph the station in advance the cause of sending the "stop and examine train" signal.

Should the signalman receiving the signal have reason to suppose that there is any danger to a train traveling in the opposite direction, he must also stop that train, and inform the engine-driver of the circumstances, instructing him to proceed cautiously. Should a train pass a signal station without a tail-lamp on the last vehicle, the signalman must not telegraph "line clear" to the station in the rear, but must call the attention of such station in the authorized manner, and on gaining attention, must give the "train passed without tail-lamp" signal. This signal having been acknowledged, the signalman at the rear station will, thereupon, stop any train following, and verbally instruct the engine-driver to proceed cautiously toward the station in advance, informing him why it is necessary that he should do so. As soon as the train, the engine-driver of which has been cautioned, has passed the signal station from whence the "train passed without tail-lamp" signal was received, the signalman there will recommence signaling in the ordinary manner.

Should any vehicle or portion of a train be running back in the wrong direction, the signalman must call the attention of the signalman at the next signal box toward which the vehicle or portion of the train may be running, by giving the prescribed signal indicating that vehicles are running back on wrong line.

The signalman who has received this signal must stop any train about to proceed on the same line, and take such protective measures as may be necessary, such as turning the runaway train across to the other line or into a siding, as may be most expedient under the circumstances.

If any vehicle or portion of a train has escaped, and is running away in the proper direction on the right line,[1] the station in advance must be advised of the fact by giving on the bell or gong the signal " vehicles running away on proper line." The signalman receiving this signal must, if necessary, send the signal forward, and take such other measures as he may consider most expedient under the circumstances.

When a train has become divided and is running on a falling gradient, the front portion must not, when the line is clear for it to proceed beyond the signals, be stopped so as to risk its being overtaken by the second portion, but when such train is running on a rising gradient, or where the line is level, the first portion must be stopped and shunted into a siding as expeditiously as circumstances will permit.

" A." In the event of any failure of the instruments or bells, so that the necessary signals can not be forwarded and received, no train must, under any circumstances, be allowed to pass a signal station into that section of the line where the failure exists, without having been previously brought to a stand, and the engine-driver and guard advised of the circumstances. When this has been done, the engine-driver must be instructed to proceed cautiously to the post in advance, so as to be able to stop short of any obstruction there may be on the line. No train must be allowed to follow

---

[1]. Not the *right hand* track. *M. M. K.*

another within five minutes; nor, when a tunnel intervenes in a block section, within ten minutes, unless the signalman on duty can satisfy himself that the tunnel is clear.[1]

Steps must be immediately taken to have the telegraphic apparatus put into working order again.

"B." To prevent delays to breakdown van trains[2] when proceeding to clear the line, they must, in all cases, be signaled as "passenger trains," the signal "shunt for fast train" being given whenever the sections in advance are occupied by trains which the breakdown gang must pass to reach the scene of accident. The same course is to be adopted in the case of an engine proceeding to take the place of one that has failed, or of an engine with or without a train, when sent forward to render assistance in cases of failure or accident to preceding trains.

Should any obstructions occur necessitating the working of single line, the person in charge, who gives the necessary instructions for so doing, must, at the same time, give written instructions for suspending the working of the

---

1. "The engine-driver must protect his engine, in accordance with the regulations, without reference to any telegraphic communications that may exist between stations or signal boxes, and he is not in any way relieved from this duty by the existence of block or other telegraphic working."—*Eng. Standard.*

2. Wrecking trains.—*M. M. K.*

line by block telegraph, " except on inclines or through tunnels, where the block telegraph working may not be suspended on special instructions being given."[1]

On the working of the double line being resumed, the order suspending the working of the line by block telegraph is to be canceled by a written notice in the same manner, and at the same time, as the order for working the single line is canceled.

Where the block system is in operation, goods, mineral, cattle, and ballast trains must be shunted out of the way of passenger trains, and mineral, slow goods, and ballast trains must also be shunted out of the way of fast goods and fish trains at stations or sidings where there are fixed signals,[2] in sufficient time to prevent the passenger train, fast goods or fish trains, respectively, being delayed by the signals either at the station where the train is being shunted or at the block station in the rear.

Where the block system is in operation, and it is necessary to foul[3] or occupy any portion of the line outside the home signal, the line must first be blocked back by telegraph to the signal box in rear before such obstruction is permitted, and during a fog or snow-storm, or

---

1. Great Wes. Ry., Eng.
2. *i. e.*, Semaphore signals, etc.—*M. M. K.*
3. Obstruct—*M. M. K.*

where, in consequence of the station being approached upon a falling gradient, special instructions for working are issued, no obstruction must be allowed at the station inside the home signal, until the line is so blocked back to the signal box in rear.

# INDEX.

|  | PAGE. |
|---|---|
| Absolute Block System, Rules for Working | 253 |
| Agents, Delivery of Freight | 207 |
| " Directions in reference to Fuel | 214 |
| " " " " " Switches | 215 |
| " " " " " Trains and Cars | 216 |
| " Freight, Miscellaneous Rules | 211 |
| " Freight Releases | 202 |
| " Freight Traffic Rules | 196 |
| " General Directions | 218 |
| " Loading and Unloading Freight | 203 |
| " Passenger Traffic Rules | 195 |
| " Receipting for Freight | 199 |
| " Receiving Freight for Shipment | 197 |
| " Sealing Freight Cars | 210 |
| " Way-billing Freight | 209 |
| Ahead of Time | 51 |
| Approaching Stations, Trains | 99 |
| Arranging Rules and Regulations, Plan Pursued in | 65 |
| Austrian Railway Regulations Governing the Passenger Service | 228 |
| Baggagemen, Train and Station | 174 |
| Behind Time | 51 |
| Bell-cord Signals | 76 |
| Bell must be Rung | 79 |
| Block System | 8, 51 |
| " " Absolute, Rules for Working | 253 |
| Blue Signals | 70, 75 |
| Brake | 52 |
| Brakemen, Freight, Rules for | 173 |
| " General Instructions to | 169 |
| " Passenger, Rules for | 171 |

## Index

| | PAGE. |
|---|---|
| Breaking in Two of Trains | 94 |
| Care Must be Exercised in Loading Freight | 205 |
| Cars | 52 |
| " Coupling | 124 |
| " Directions to Agents in reference to | 216 |
| " Sealing of | 210 |
| Classes of Trains | 53, 81 |
| Clearing a Train | 53 |
| Closed Switch | 53 |
| Collection of Fares on English Roads | 12 |
| Compensation paid " " " | 243 |
| Compiling Rules and Regulations, Plan pursued in | 65 |
| Conditions of Service on English Roads | 243 |
| Conductor, Diversity of Duties Abroad | 13 |
| Conductors, Freight, Rules for | 159 |
| " General Instructions to | 147 |
| " Passenger, Rules for | 154 |
| " Signals by Bell-cord | 76 |
| Conservatism of Trainmen | 36 |
| Construction and Wood Trains | 111 |
| Construction Train | 53 |
| Coupling Cars | 124 |
| Danger of Dissimilarity of Signals | 27 |
| Delayed Trains | 104 |
| Delivery of Freight | 207 |
| Differences Observable in Rules | 38 |
| Discrimination Exercised by Trainmen | 34 |
| Dispatcher, Train | 61 |
| Dissimilarity of Signals in use | 27 |
| Double Track Lines, Rules for | 117 |
| Engine Bell must be Rung | 79 |
| Engine Inspectors, Rules for | 184 |
| Enginemen, Rules for | 174 |
| Enginemen's Signals | 75 |
| Engine Supplies | 177 |
| English Roads, Absolute Block System | 253 |
| " " Compensation | 243 |
| " " Conditions of Service | 243 |

## Index.

|  | PAGE. |
|---|---|
| English Roads, Manipulation of Trains upon | 12 |
| " " Regulations of | 238 |
| " " Security Required | 243 |
| " " Uniforms Required | 251 |
| Explanation of Terms | 51 |
| Extra Trains | 54, 81, 106 |
| Fares, Collection of, on English Roads | 12 |
| Firemen, Rules for | 182 |
| Flags to be used as Signals | 69 |
| Flying Switch | 54 |
| Following other Trains | 101 |
| Force Employed upon English Roads | 12 |
| Four Tracks, Use and Value of | 7 |
| Freight Agents, Miscellaneous Rules for | 211 |
| " Brakemen, Rules for | 173 |
| " Care must be exercised in Loading | 205 |
| " Cars, Sealing of | 210 |
| " Conductors, Rules for | 159 |
| " Delivery of | 207 |
| " From and To Stations at which there are no Agents | 208 |
| " Loading and Unloading | 203 |
| " Receipting for | 199 |
| " Receiving for Shipment | 197 |
| " Traffic Regulations | 196 |
| " Trains, Kinds of | 81 |
| " Way-Billing | 209 |
| Fuel, Directions to Agents in Reference to | 214 |
| Fusee Signals. | 71 |
| Ganger | 49, 50 |
| General Instructions | 221 |
| General Instructions to Agents | 218 |
| " " " Brakemen | 169 |
| " " " Conductors | 147 |
| General Regulations for Working Block System | 253 |
| Grade of Trains | 54, 81 |
| Green and White Signals | 70, 79 |
| Green Signals | 69, 74 |

## Index.

|  | PAGE. |
|---|---|
| Hand Signals | 77 |
| Holding a Train | 54 |
| Individuality of Railroad Companies | 27 |
| Inspectors of Engines, Rules for | 184 |
| Instructions, General | 221 |
| Instructions to Agents in reference to Delivery of Freight | 207 |
| " Agents in reference to Freight Releases | 202 |
| " Agents in reference to Freight Traffic | 196 |
| " Agents in reference to Fuel | 214 |
| " Agents in reference to Loading and Unloading Freight | 203 |
| " Agents in reference to Passenger Traffic | 195 |
| " Agents in reference to Receipting for Freight | 199 |
| " Agents in reference to Receiving Freight | 197 |
| " Agents in reference to Sealing Freight Cars | 210 |
| " Agents in reference to Switches | 215 |
| " Agents in reference to Trains and Cars | 216 |
| " Agents in reference to Waybilling Freight | 209 |
| " Brakemen, General | 169 |
| " Conductors, General | 147 |
| " Engine Inspectors | 184 |
| " Enginemen | 174 |
| " Firemen | 182 |
| " Freight Brakemen | 173 |
| " Freight Conductors | 159 |
| " Passenger Brakemen | 171 |
| " Passenger Conductors | 154 |
| " Telegraph Operators | 188 |
| " Telegraph Repairers | 193 |
| " Trackmen | 129 |
| Intelligent Discrimination Exercised by Trainmen | 34 |
| Irregular Train | 55 |
| Keep off Time of a Train | 55, 103 |
| L, Rule, for Protection of Trains | 85 |
| Lack of Completeness in Framing Rules | 43 |
| Lamp Signals | 69, 78 |
| Lay Bye | 49 |

## Index. 267

| | PAGE. |
|---|---|
| Loading and Unloading Freight | 203 |
| Loading Freight, Care must be exercised in | 205 |
| Lorry | 49, 50 |
| Lost its Rights | 55 |
| Lost Time | 55 |
| Main Track | 55 |
| Making Time | 55 |
| Manipulation of Trains upon English Roads | 12 |
| Meeting or Passing Trains | 97 |
| Meeting Point | 55 |
| Middle Sidings | 123 |
| Miscellaneous Rules for Freight Agents | 211 |
| Miscellaneous Train Orders | 125 |
| Movement of Trains by Telegraphic Orders | 18, 55, 139 |
| Must Stop, Trains | 97 |
| Mysteries that underlie the Organization and Movement of Trains | 1 |
| On Time | 55 |
| Open Switch | 56 |
| Operators, Telegraph, Rules for | 188 |
| Organization of Trains | 1 |
| Overshooting | 56 |
| Passenger Brakemen, Rules for | 171 |
|     " Conductors, Rules for | 154 |
|     " Service, Austrian Regulations | 228 |
|     " Traffic, Rules for Agents | 195 |
|     " Trains, Kinds of | 81 |
| Passing Other Trains | 97 |
| Passing Point | 56 |
| Phraseology of Trainmen | 46 |
|     " peculiar to English Roads | 49 |
| Plan Pursued in Arranging and Compiling these Rules and Regulations | 65 |
| Protection of Trains | 23 |
| Protection of Trains, Rule L | 85 |
| Railroad Companies, Individuality of | 27 |
| Receipting for Freight | 199 |
| Receiving Freight for Shipment | 197 |

|   |   | PAGE |
|---|---|---|
| Red Signals | | 69, 72, 73, 80 |
| Regular Trains | | 56, 81 |
| Regulations Governing Use of Signals | | 78 |
| " Lack of Completeness in Framing | | 43 |
| " of Austrian Railways Governing Passenger Service | | 228 |
| " of English Roads | | 238 |
| " Partake of the Character of the men Introducing them | | 37 |
| Releases | | 202 |
| Repairers, Telegraph, Rules for | | 193 |
| Rights of a Train | | 56, 82 |
| Right to the Road | | 56 |
| Rule L. for Protection of Trains | | 85 |
| Rules, Lack of Completeness in Framing | | 43 |
| Rules of the Great English Roads | | 238 |
| Running Against a Train | | 57 |
| Running Time of Trains | | 57 |
| Running with Care | | 96 |
| Run Regardless | | 57 |
| Salaries Paid in England | | 243 |
| Schedule by which Trains are Operated | | 15, 57 |
| Scotch Block | | 49 |
| Sealing Freight Cars | | 210 |
| Sectionmen, Rules for | | 129 |
| Security Required from Employés in England | | 243 |
| Semaphore | | 58 |
| Setting a Switch | | 58 |
| Shunting | | 58 |
| Side Track | | 58 |
| Sidings (see Side Track). | | |
| Signals | | 59, 69 |
| Signals, Blue | | 70, 75 |
| " Conductor's Bell Cord | | 76 |
| " Danger of Dissimilarity in | | 27 |
| " Enginemen's | | 75 |
| " Fusees as | | 71 |
| " Green | | 69, 74 |

## Index. 269

|  | PAGE. |
|---|---|
| Signals, Green and White | 70, 79 |
| " Hand | 77 |
| " in Use, Dissimilarity of | 27 |
| " Red | 69, 72, 73, 80 |
| " Regulations Governing Use | 78 |
| " Required by Railway Companies | 69 |
| " Semaphore | 72 |
| " Switch | 80 |
| " Torpedoes | 70, 85 |
| " Train | 72 |
| " Whistle | 75, 76 |
| " White | 69, 70, 74, 126 |
| " Yellow | 75 |
| Single Track, Skill Required to Move Trains upon | 6 |
| Slipping the Wheels | 59 |
| Some of the Differences in Rules | 38 |
| Special Train | 59 |
| Speed of Trains | 116 |
| Spur Track | 60 |
| Station | 60 |
| Station Baggagemen | 174 |
| Supplies, Engine | 177 |
| Supplies, Train | 149 |
| Switch | 60 |
| Switch Signals | 80 |
| Switches, Directions to Agents in Reference to | 215 |
| Switching | 61 |
| Technical Terms, Explanation of | 51 |
| Telegraph Department, Want of Uniformity in | 42 |
| Telegraphic Orders, Movement of Trains by | 18, 139 |
| Telegraph Operators, Rules for | 188 |
| " Repairers, Rules for | 193 |
| Terms in Use, Explanation of | 51 |
| Third Track | 61, 123 |
| Three Tracks, Use and Value of | 6 |
| Through Trains | 61 |
| Time | 61 |
| Time-Table | 15, 57 |

|   | PAGE. |
|---|---|
| Torpedoes, Signals, | 70, 71, 85 |
| Track, the | 129 |
| Trackmen Rules for | 129 |
| Traffic, Freight, Rules for Agents | 196 |
| " Passenger, Rules for Agents | 195 |
| Train and Station Baggagemen | 174 |
| Train Dispatcher | 61 |
| Train Signals | 72 |
| Train Staff | 49, 50 |
| Train Supplies | 149 |
| Trainmen, Conservatism of | 36 |
| " Intelligent Discrimination Exercised by | 34 |
| Trains | 62 |
| " Approaching Stations | 99 |
| " Breaking in Two | 94 |
| " Construction and Wood | 111 |
| " Delayed | 104 |
| " Directions to Agents in Reference to | 216 |
| " Extra | 54, 81, 106 |
| " Following Other Trains | 101 |
| " Keeping off Time of Other Trains | 103 |
| " Meeting or Passing | 97 |
| " Miscellaneous Orders Relative to | 125 |
| " Moved by Telegraph | 139 |
| " Movement of by Telegraphic Orders | 18 |
| " Must Stop | 97 |
| " Protection of | 23, 85 |
| " Rights of | 56, 82 |
| " Running with Care | 96 |
| " Speed of | 116 |
| " Upon English Roads, Manipulation of | 12 |
| " Wild | 64, 81, 115 |
| Trolley | 49 |
| Turn a Switch | 62 |
| Turn-Out or Side-Track | 58 |
| Two Track Line, Rules for | 117 |
| Uniformity in Rules, Want of | 32 |
| Uniforms Worn on English Roads | 251 |

Index.

|  | PAGE. |
|---|---|
| Unloading Freight | 203 |
| Use of Signals, Regulations Governing | 78 |
| Value of Four Tracks | 7 |
| " Three " | 6 |
| " Two " | 6 |
| Wages Paid in England | 243 |
| Want of Completeness in Framing Rules | 43 |
| " Uniformity in Rules | 32 |
| " " " in Telegraph Department | 42 |
| Way Bill | 62 |
| Way-billing Freight | 209 |
| Way Train | 62 |
| When a Train has Lost its Rights | 63 |
| When Trains Break in Two | 94 |
| Whistle Signals | 75, 76 |
| Whistling-Post | 64 |
| White Signals | 69, 70, 74, 126 |
| Wild Train | 64, 81, 115 |
| Wonderful Phraseology of Trainmen | 46 |
| Wood Trains | 64, 111 |
| Y | 64 |
| Yardmaster, Rules for | 185 |
| Yellow Signals | 75 |

www.ingramcontent.com/pod-product-compliance
Lightning Source LLC
Chambersburg PA
CBHW031250250426
43672CB00029BA/1916